# Unification of the *Eleven* Boson Interactions based on 'Rotations of Interactions'

## New SU(3)⊗U(64) Symmetry and 'Interactions of Interactions,' ElectroWeak Theory as an Interaction Rotation, Gravity: Earth Scale, Galactic Scale, Intergalactic Scale, Proton Radius and Spin Puzzles Resolved

***STEPHEN BLAHA***
***BLAHA RESEARCH***

***Pingree Hill Publishing***

ISBN: 978-0-9984575-3-6



**Cover Credits**



Rev. 00/00/01 January 15, 2017

## Some Other Books by Stephen Blaha

*All the Megaverse! Starships Exploring the Endless Universes of the Cosmos using the Baryonic Force* (Blaha Research, Auburn, NH, 2014)

*SuperCivilizations: Civilizations as Superorganisms* (McMann-Fisher Publishing, Auburn, NH, 2010)

*Universes and Megaverses: From a New Standard Model to a Physical Megaverse; The Big Bang; Our Sister Universe's Wormhole; Origin of the Cosmological Constant, Spatial Asymmetry of the Universe, and its Web of Galaxies; A Baryonic Field between Universes and Particles; Flatverse Extended Wheeler-DeWitt Equation* (Blaha Research, Auburn, NH, 2014)

*PHYSICS IS LOGIC PAINTED ON THE VOID: Origin of Bare Masses and The Standard Model in Logic, U(4) Origin of the Generations, Normal and Dark Baryonic Forces, Dark Matter, Dark Energy, The Big Bang, Complex General Relativity, A Megaverse of Universe Particles* (Blaha Research, Auburn, NH, 2015).

*PHYSICS IS LOGIC Part II: The Theory of Everything, The Megaverse Theory of Everything, U(4)⊗U(4) Grand Unified Theory (GUT), Inertial Mass = Gravitational Mass, Unified Extended Standard Model and a New Complex General Relativity with Higgs Particles, Generation Group Higgs Particles* (Blaha Research, Auburn, NH, 2015).

*The Origin of Higgs ("God") Particles and the Higgs Mechanism: Physics is Logic III, Beyond Higgs – A Revamped Theory With a Local Arrow of Time, The Theory of Everything Enhanced, Why Inertial Frames are Special, Universes of the Mind* (Blaha Research, Auburn, NH, 2015).

*The Origin of the Eight Coupling Constants of The Theory of Everything: U(8) Grand Unified Theory of Everything (GUTE), $S^8$ Coupling Constant Symmetry, Space-Time Dependent Coupling Constants, Big Bang Vacuum Coupling Constants, Physics is Logic IV* (Blaha Research, Auburn, NH, 2015).

*New Types of Dark Matter, Big Bang Equipartition, and A New U(4) Symmetry in the Theory of Everything: Equipartition Principle for Fermions, Matter is 83.33% Dark, Penetrating the Veil of the Big Bang, Explicit QFT Quark Confinement and Charmonium, Physics is Logic V* (Blaha Research, Auburn, NH, 2015).

*The Periodic Table of the 192 Quarks and Leptons in The Theory of Everything: The U(4) Layer Group, Physics is Logic VI* (Blaha Research, Auburn, NH, 2015).

*New Boson Quantum Field Theory, Dark Matter Dynamics, Dark Matter Fermion Layer Mixing, Genesis of Higgs Particles, New Layer Higgs Masses, Higgs Coupling Constants, Non-Abelian Higgs Gauge Fields, Physics is Logic VII* (Blaha Research, Auburn, NH, 2015)

*Unification of the Strong Interactions and Gravitation: Quark Confinement Linked to Modified Short-Distance Gravity; Physics is Logic VIII* (Blaha Research, Auburn, NH, 2016).

*Unification of the Seven Boson Interactions based on the Riemann-Christoffel Curvature Tensor* (Pingree Hill Publishing, Auburn, NH, 2016).

*CQMechanics: A Unification of Quantum & Classical Mechanics, Quantum/Semi-Classical Entanglement, Quantum/Classical Path Integrals, Quantum/Classical Chaos* (Blaha Research, Auburn, NH, 2016).

Available on Amazon.com, Amazon.co.uk, bn.com, and other international web sites as well as at better bookstores (through Ingram Distributors).

# Preface

This book is a supplement to *Unification of the Seven Boson Interactions based on the Riemann-Christoffel Curvature Tensor ...* (I). Together these works provide a unified picture of the boson sector of The Theory of Everything: vector bosons, gravitation and Higgs bosons. The boson sector unification is based on the Riemann-Christoffel tensor, which must include all interactions in its description of our curved space-time. The boson lagrangian part of The Theory of Everything is constructed from the Riemann-Christoffel tensor in a manner similar to the construction of the Theory of Gravity. In I we constructed a unified theory for the seven necessarily known interactions required by experimental data. In this book we add four interactions that would also seem to be needed: a gravitational vector spinor interaction, an SU(2) Dark Weak interaction a U(1) Dark Weak interaction, and a 'unifying' interaction that we call the Omega-Interaction. We attribute Dark Weak interactions to Dark Matter as a natural extension of the set of particle interactions. We do not introduce a Dark Strong interaction due to the cosmological data that shows Dark Matter to be diffused in lumps throughout the universe and apparently not aggregated into stars and planets. This difference suggests that a Dark nuclear force and consequently a Dark Strong interaction does not exist. After assembling the eleven forces in the Riemann-Christoffel tensor, we note that a (broken) SU(3) by U(64) symmetry is manifest and we develop the consequences of this symmetry. We take this symmetry to be a local Yang-Mills symmetry and apply the Faddeev-Popov Mechanism to generate 'ghost' interactions. Thus we go beyond the Standard Model in two ways: interactions between interactions, and ghost interactions between interactions. We conclude by showing that the gravity potential found at solar system distances, at galactic distances, and at inter-galactic distances follow from our theory. We also show that the theory yields a linear potential and the Charmonium potential. Lastly we show the theory may explain the missing proton spin, and also the difference in proton radius found in hydrogen and muonic hydrogen.

# CONTENTS

# 1. Unification of The Eleven Interactions

In Blaha (2016h), which we denote I in this Supplement, we established a unified theory of gauge boson interactions through the use of the Riemann-Christoffel tensor. The justification for using the Riemann-Christoffel tensor is the simple fact that the curvature of space-time depends on the energy-momentum of particles within it. We used this justification to describe the unification of seven known interactions – known because they are necessary to understand the structure and contents of the set of known fermions (with the addition of a fourth generation of fermions which the author anticipates.)

In this Supplement we add four additional gauge field interactions to create a set of eleven interactions that would appear to be complete unless a radically new set of interactions and particles should be found.

The set of gauge boson interactions that we now consider are:

Vector Gauge Interactions
The SU(2) Weak interaction
The U(1) Electromagnetic interaction
The SU(3) Strong interaction
The U(4) Generation group interaction
The U(4) Layer group interaction
The SU(2) Dark Weak interaction
The U(1) Dark Electromagnetic interaction
The U(4) General Relativistic Reality group interaction
The SU(3)⊗U(64) Interaction Rotation group Interaction
The spinor connection interaction

Spin 2 Gauge Interaction
The Gravitational field

We shall see that we can unify these interactions within the framework of the general covariant derivative and the Riemann-Christoffel tensor to achieve a unified theory of bosonic interactions. We will then consider U(64) rotations of the field equations of both the bosonic and fermionic sectors. Although this rotation 'symmetry' is clearly broken, we can use the Faddeev-Popov Mechanism to implement gauge conditions with significant consequences. These conditions implemented within a Path Integral formulation lead to a new class of 'ghost' interactions amongst eight rotate-able gauge interactions that go beyond those of The Standard Model.

Together with the 'interactions between interactions' that we found in I we have new mechanisms to account for deviations from conventional Standard Model theory. Among the possible problems they resolve are: 1) the deviations from G/r of the gravitational potential within galaxies (MoND), and additional deviations from G/r between galaxies, 2) a possible explanation of the proton spin puzzle due to a new gluon-photon contact interaction, and 3) a possible explanation of proton radius measurement discrepancies due to the Generation group gauge field – photon interactions.

# 2. The Eleven Interactions

In this chapter we define the eleven interactions of our unified boson theory.[1] We begin by defining the fields of the eleven interactions using the pseudoquantization formalism described in Blaha (2016c) and earlier books. We summarize the description of the seven interactions found in I plus four more interactions. We use the Pseudoquantization formalism described in I that defines two gauge fields for each gauge interaction.

## 2.1 Concept of the Eleven Interaction Gauge Fields

This section defines the eleven interactions. The seven interactions in I are clearly needed interactions, in the author's view, to account for the Standard Model interactions and the form of the fermion 'periodic table' with three provisos: 1) we assume a 4th generation of fermions which leads to a broken U(4) group called the Generation group, 2) we assume another U(4) group called the Layer group (based on certain number conservation laws for generations that imply another U(4) group), and 3) we assume the universe is described by complex coordinate systems that yield another U(4) symmetry group called the General Coordinate Reality group.

We add SU(2) and U(1) group interactions for Dark Matter in analogy with corresponding normal matter interactions. We include the spinor connection interaction. And we introduce an interaction, called the *Ω-interaction*, that 'rotates' the other interactions. Thus this supplement includes four additional interactions. We believe that the total of eleven interactions completes the roster of interactions involving particles that we know of, or anticipate, including normal and Dark matter.

## 2.2 Summary of the Eleven Interactions

### 2.2.1 The SU(2) Weak Interaction

The Weak interaction SU(2) gauge fields are defined as $W^{1i\mu}(x)$ and $W^{2i\mu}(x)$ for i = 1, 2, 3.[2] Using SU(2) generators we define the matrix form by $W^{k\mu}(x) = W^{ki\mu}(x)\tau_i$ for k= 1, 2. Under an SU(2) gauge transformation $C_W$ the gauge fields transform as

$$W^{1\mu}(x) \rightarrow C_W(x)W^{1\mu}(x)C_W^{-1}(x) - i\, C_W(x)\partial^\mu C_W^{-1}(x) \qquad (2.1)$$

and

[1] The fermionic sector will be considered later.

[2] We define two fields for each interaction using the Pseudoquantization formalism that we developed in earlier books.

$$W^{2\mu}(x) \rightarrow C_W(x)W^{2\mu}(x)C_W^{-1}(x) \qquad (2.2)$$

### 2.2.2 The U(1) Electromagnetic Interaction

The U(1) electromagnetic gauge fields[3] are defined as $A_E^{1\mu}(x)$ and $A_E^{2\mu}(x)$. Under a local electromagnetic gauge transformation $C_E(x)$ the gauge fields transform as

$$A_E^{1\mu}(x) \rightarrow C_E(x)A_E^{1\mu}(x)C_E^{-1}(x) - i\, C_E(x)\partial^\mu C_E^{-1}(x) \qquad (2.3)$$

and

$$A_E^{2\mu}(x) \rightarrow C_E(x)A_E^{2\mu}(x)C_E^{-1}(x) \qquad (2.4)$$

### 2.2.3 The SU(3) Strong Interaction

The Strong SU(3) gauge fields are defined as $A_{SU(3)}{}^{1i\mu}(x)$ and $A_{SU(3)}{}^{2i\mu}(x)$ for i = 1, … , 8. Using SU(3) generators we define the matrix form by $A_{SU(3)}{}^{1\mu}(x) = A_{SU(3)}{}^{1i\mu}(x)T_i$ and $A_{SU(3)}{}^{2\mu}(x) = A_{SU(3)}{}^{2i\mu}(x)T_i$. Under an SU(3) gauge transformation C the gauge field transforms as

$$A_{SU(3)}{}^{1\mu}(x) \rightarrow C(x)A_{SU(3)}{}^{1\mu}(x)C^{-1}(x) - i\, C(x)\partial^\mu C^{-1}(x) \qquad (2.5)$$

and

$$A_{SU(3)}{}^{2\mu}(x) \rightarrow C(x)A_{SU(3)}{}^{2\mu}(x)C^{-1}(x) \qquad (2.6)$$

### 2.2.4 The U(4) Generation Group Interaction

The U(4) Generation group[4] generators are denoted $G_i$ and its gauge fields are denoted $U_{\mu i}(X)$. Thus the Generation group terms in covariant derivatives are

$$g_G \mathbf{U}^i{}_\mu \cdot \mathbf{G} \qquad (2.7)$$

where $g_G$ is the coupling constant, and i = 1, 2. Under a U(4) Generation transformation $C_G$ the gauge field transforms as

$$U^{1\mu}(x) \rightarrow C_G(x)U^{1\mu}(x)C_G^{-1}(x) - i\, C_G(x)\partial^\mu C_G^{-1}(x) \qquad (2.8)$$

and

$$U^{2\mu}(x) \rightarrow C_G(x)U^{2\mu}(x)C_G^{-1}(x) \qquad (2.9)$$

### 2.2.5 The U(4) Layer Group Interaction

The U(4) Layer group[5] generators are denoted $G_{Lk}$ and its gauge fields are denoted $V^i{}_{\mu k}(X)$. Thus the Layer group terms in covariant derivatives are

---

[3] We introduce two fields as we did in our article S. Blaha, Phys. Rev. D**10**, 4268 (July, 1974). These fields enable us to define a free electromagnetic lagrangian that is linear in the fields for reasons given elsewhere.
[4] If there are only three generatons of fermions then the Generation group is U(3).
[5] If there are only three generatons of fermions then the Layer group is also U(3).

$$g_V \mathbf{V}^i{}_\mu \cdot \mathbf{G}_L \qquad (2.10)$$

where $g_V$ is the coupling constant, i = 1, 2, and k = 1, 2, …, 16. Under a U(4) Generation transformation $C_G$ the gauge field transforms as

$$V^{1\mu}(x) \rightarrow C_G(x)V^{1\mu}(x)C_G^{-1}(x) - i\, C_G(x)\partial^\mu C_G^{-1}(x) \qquad (2.11)$$

and

$$V^{2\mu}(x) \rightarrow C_G(x)V^{2\mu}(x)C_G^{-1}(x) \qquad (2.12)$$

### 2.2.6 The SU(2) Dark Weak Interaction

We assume Dark Weak interactions have the same form as the known SU(2) Weak interactions. the The Dark Weak interaction SU(2) gauge fields are defined as $W_D^{1i\mu}(x)$ and $W_D^{2i\mu}(x)$ for i = 1, 2, 3. Using SU(2) generators we define the matrix form by $W_D^{k\mu}(x) = W_D^{ki\mu}(x)\tau_{Di}$ for k= 1, 2 where the generator matrices $\tau_{Di}$ are not in the same subspace as the normal SU(2) generators. Under a Dark SU(2) gauge transformation $C_{DW}$ the gauge field transforms as

$$W_D^{1\mu}(x) \rightarrow C_{DW}(x)W_D^{1\mu}(x)C_{DW}^{-1}(x) - i\, C_{DW}(x)\partial^\mu C_{DW}^{-1}(x) \qquad (2.13)$$

and

$$W_D^{2\mu}(x) \rightarrow C_{DW}(x)W_D^{2\mu}(x)C_{DW}^{-1}(x) \qquad (2.14)$$

### 2.2.7 The U(1) Dark Electromagnetic Interaction

The U(1) Dark electromagnetic gauge field[6] are defined as $A_{DE}^{1\mu}(x)$ and $A_{DE}^{2\mu}(x)$. Under a local Dark electromagnetic gauge transformation $C_{DE}(x)$ the gauge fields transform as

$$A_{DE}^{1\mu}(x) \rightarrow C_{DE}(x)A_{DE}^{1\mu}(x)C_{DE}^{-1}(x) - i\, C_{DE}(x)\partial^\mu C_{DE}^{-1}(x) \qquad (2.15)$$

and

$$A_{DE}^{2\mu}(x) \rightarrow C_{DE}(x)A_{DE}^{2\mu}(x)C_{DE}^{-1}(x) \qquad (2.16)$$

### 2.2.8 The U(4) General Relativistic Reality Group Interaction

As shown in I we can factor complex General Coordinate transformations into a product of a complex General coordinate transformation, which can be expressed in terms of a U(4) gauge field, and a real-valued General Coordinate transformation. We call the U(4) gauge field the General Relativistic Reality Group interaction gauge field. We denote this gauge field and its secondary gauge field as $A_{Rflat}^{1\mu}(x)$ and $A_{Rflat}^{2\mu}(x)$. Under a gauge transformation $C_R(x)$ they transform as

$$A_{Rflat}^{1\mu}(x) \rightarrow C_R(x)A_{Rflat}^{1\mu}(x)C_R^{-1}(x) - i\, C_R(x)\partial^\mu C_R^{-1}(x) \qquad (2.17)$$
$$A_{Rflat}^{2\mu}(x) \rightarrow C_R(x)A_{Rflat}^{2\mu}(x)C_R^{-1}(x)$$

---

[6] We introduce two fields as we did in our article S. Blaha, Phys. Rev. D**10**, 4268 (July, 1974). These fields enable us to define a free electromagnetic lagrangian that is linear in the fields for reasons given elsewhere.

## 2.2.9 The 'Interaction Rotation' Interaction - $A_\Omega$

The SU3)⊗U(64) interaction rotation group gauge fields[7] are defined as $A_\Omega{}^{1ij\mu}(x)$ and $A_\Omega{}^{2ij\mu}(x)$ for i = 1, …, 8 and j = 1, …, 64. They total to 192 gauge field components. Using their 72 generators expressed in the SU(3) 3 representation and the U(64) 64 representation, with matrix denoted $T_{\Omega ij}$, we can define the matrix form as

$$A_\Omega{}^{k\mu}(x) = A_\Omega{}^{kij\mu}(x)T_{\Omega ij} \tag{2.18}$$

for k = 1, 2, where $T_{\Omega ij}$ is a product of SU3)⊗U(64) generators. The tensor product generator matrices are 192×192 matrices. We choose 192×192 matrics due to the 192 fermions in our Periodic Table of Fermions (chapter 3.)[8]

Under a local SU3)⊗U(64) gauge transformation $C_\Omega$ a gauge field transforms as

$$A_\Omega{}^{1\mu}(x) \rightarrow C_\Omega(x)A_\Omega{}^{1\mu}(x)C_\Omega{}^{-1}(x) - i\, C_\Omega(x)\partial^\mu C_\Omega{}^{-1}(x) \tag{2.19}$$
$$A_\Omega{}^{2\mu}(x) \rightarrow C_\Omega(x)A_\Omega{}^{2\mu}(x)C_\Omega{}^{-1}(x)$$

This interaction, which we will call the *Ω-interaction,* is described in detail in chapter 3. The gauge fields will correspondingly be called *Ω-fields.*

## 2.2.10 The Spinor Connection Interaction

The spinor connection used in formulations of vierbein gravity is $B^1{}_{\mu ab}(x)$ where a and b are tangent space indices. The vector is combined with γ matrices for use in matrix equations:

$$B^{1\mu} = B^{1\mu}{}_{ab}\Sigma^{ab} \tag{2.20}$$

where

$$\Sigma^{ab} = i\,[\gamma^a, \gamma^b]/4 \tag{2.21}$$

Under a local Lorentz transformation S

$$B^{1\mu}(x) \rightarrow S(x)B^{1\mu}(x)S^{-1}(x) - i\, S(x)\partial^\mu S^{-1}(x) \tag{2.22}$$

Similarly we define a secondary spinor connection:

$$B^{2\mu} = B^{2\mu}{}_{ab}\Sigma^{ab} \tag{2.23}$$

where

$$\Sigma^{ab} = i\,[\gamma^a, \gamma^b]/4 \tag{2.24}$$

Under a local Lorentz transformation S

---

[7] The SU(3) factor is *not* color SU(3).

[8] See pp. 388-389 in M. Hamermesh, Group Theory (Addison-Wesley, New York, 1962) for the definition of U(n) generator matrices.

$$B^{2\mu}(x) \rightarrow S(x)B^{2\mu}(x)S^{-1}(x) \tag{2.25}$$

A simple spin ½ field transforms as

$$(\partial^{\mu} + i\,B^{1\mu} + i\,B^{2\mu})\psi \rightarrow S(\partial^{\mu} + i\,B^{1\mu} + i\,B^{2\mu})\psi \tag{2.26}$$

### 2.2.11 Real-Valued General Coordinate Connection (Interaction)

The usual gravitational metric field $g_{\mu\nu}$ is supplemented with a secondary metric field to enable us to define a higher derivative gravitation theory and still use the canonical Euler-Lagrange formalism to generate dynamical equations. Thus we define a second metric field and associated affine connection

$$g^{2\mu\nu} = g^{2\nu\mu} \tag{2.27}$$

$$\Gamma_{GR}{}^{2\lambda}{}_{\mu\nu} = \tfrac{1}{2}g^{2\lambda\alpha}(\partial_{\mu}g^{2}{}_{\alpha\nu} + \partial_{\nu}g^{2}{}_{\alpha\mu} - \partial_{\alpha}g^{2}{}_{\mu\nu}) \tag{2.28}$$

in addition to the usual real-valued metric and affine connection denoted $\Gamma_{GR}{}^{1\sigma}{}_{\nu\mu} = \Gamma_{GR}{}^{\sigma}{}_{\nu\mu}$ and the secondary connection $\Gamma_{GR}{}^{2\sigma}{}_{\nu\mu}$.

# 3. The 'Interaction Rotation' Interaction $A_\Omega$

## 3.1 $A_\Omega$ Formalism

The plenitude of interactions that we have identified in I and summarized in chapter 2 leads us to consider the possibility of a unification principle –not one of the three described in chapter 1 of I – but a deeper principle that actually leads to the third unification principle in I (unification based on the Riemann-Christoffel tensor).

This new unification principle is based on the rotation of interactions. We suggest that the set of eight particle interactions can be rotated amongst each other by local SU3)⊗U(64) transformations, which we call *Ω-transformations*. Since there are 64 fields in the eight interactions listed in sections 2.2.1 through 2.2.8 the transformations will have the form of the 1 representation of SU(3) and the 64 representation of U(64). The Ω-transformation SU3)⊗U(64) fields symmetry must be a broken symmetry. Its gauge fields acquire a mass – presumably due to the Higgs Mechanism.

We begin by defining the 8-vectors:

$$\mathbf{A}_I{}^{1\mu}(x) = (g_1\mathbf{A}_{SU(3)}{}^{1\mu}(x), g_2\mathbf{W}^{1\mu}(x), g_3\mathbf{A}_E{}^{1\mu}(x), g_4\mathbf{W}_D{}^{1\mu}(x), g_5\mathbf{A}_{DE}{}^{1\mu}(x), g_6\mathbf{U}^{1\mu}(x), g_7\mathbf{V}^{1\mu}(x), g_8\mathbf{A}_{Rflat}{}^{1\mu}(x))$$
$$\mathbf{A}_I{}^{2}{}_I{}^{\mu}(x) = (g_1\mathbf{A}_{SU(3)}{}^{2\mu}(x), g_2\mathbf{W}^{2\mu}(x), g_3\mathbf{A}_E{}^{2\mu}(x), g_4\mathbf{W}_D{}^{2\mu}(x), g_5\mathbf{A}_{DE}{}^{2\mu}(x), g_6\mathbf{U}^{2\mu}(x), g_7\mathbf{V}^{2\mu}(x), g_8\mathbf{A}_{Rflat}{}^{2\mu}(x)) \quad (3.1)$$

where each element is a vector of the gauge fields in the group of the gauge field and the respective coupling constants are labeled $g_1, g_2, \ldots, g_8$.

The interactions' symmetry is SU(3)⊗SU(2)⊗U(1)⊗SU(2)⊗U(1)⊗U(4)⊗U(4)⊗U(4) respectively. The number of fields for each of the elements of the vectors is 8. 3, 1, 3, 1, 16, 16, and 16 respectively – totaling 64 fields.

Similarly we define an 8-vector of 64 generators

$$\mathbf{T}_I = (\mathbf{T}_{SU(3)}, \boldsymbol{\tau}_{SU(2)}, \mathbf{I}_{U(1)}, \boldsymbol{\tau}_{DSU(2)}, \mathbf{I}_{DU(1)}, \mathbf{G}_{U(4)}, G_{LU(4)}, \mathbf{G}_{RflatU(4)}) \quad (3.2)$$

Then the total gauge fields interaction term within a covariant derivative corresponding to the eight interactions, the spinor interaction, and the Ω-interaction can be expressed as

$$A_\Omega{}^{1\mu}(x) + A_\Omega{}^{2\mu}(x) + \mathbf{A}_I{}^{1\mu}(x)\cdot\mathbf{T}_I + \mathbf{A}_I{}^{2\mu}(x)\cdot\mathbf{T}_I + B^{1\mu} + B^{2\mu} \quad (3.3)$$

The remaining additional interaction is the real-valued gravitational connection as we will see in chapter 4.

## 3.2 New Notation for the Eight Interactions

We now number the elements of the field vectors in eqs. 3.1 and 3.2:

$$\mathbf{A}_I^{1\mu}(x) = (g_1A_1^{1\mu}(x),\ g_2A_2^{1\mu}(x),\ g_3A_3^{1\mu}(x),\ g_4A_4^{1\mu}(x),\ g_5A_5^{1\mu}(x),\ g_6A_6^{1\mu}(x),\ g_7A_7^{1\mu}(x),\ g_8A_8^{1\mu}(x)) \qquad (3.1a)$$

$$\mathbf{A}_I^{2\mu}(x) = (g_1A_1^{2\mu}(x),\ g_2A_2^{2\mu}(x),\ g_3A_3^{2\mu}(x),\ g_4A_4^{2\mu}(x),\ g_5A_5^{2\mu}(x),\ g_6A_6^{2\mu}(x),\ g_7A_7^{2\mu}(x),\ g_8A_8^{2\mu}(x)) \qquad (3.1b)$$

respectively and the eight sets of generator matrices (totaling 64 matrices):

$$\mathbf{T}_I = (T_1, T_2, T_3, T_4, T_5, T_6, T_7, T_8) \qquad (3.2a)$$

The form of the i[th] generator can be chosen to be a diagonal 192×192 matrix with repeating blocks. For example we can choose the 3[rd] generator $T_3$ of color SU(3) will have 32 3×3 $T_3$ blocks along its diagonal because there are 32 normal color quark triplets in the set of 192 fermions.[9]

$$\mathbf{T}_{Ii} = \begin{bmatrix} \cdot & & & & & & \\ & \cdot & & & & & \\ & & T_3 & & & & \\ & & & \cdot & & & \\ & & & & T_3 & & \\ & & & & & \cdot & \\ & & & & & & \cdot \end{bmatrix}$$

Thus the Dirac equation (next section) for the 192 fermions can be constructed using 192×192 representations of the 72 generators.

## 3.3 'Interaction Rotations' for Fermions

We define a column vector $\psi$ containing the spinors of all fundamental fermions which number 192 normal and Dark fermions in our theory based on four generations in four layers as detailed in Blaha (2016a), (2016b) and (2016c). Then their Dirac equation has the form:

$$\{\partial^\mu + i\,[g_\Omega A_\Omega^{1\mu}(x) + g_\Omega A_\Omega^{2\mu}(x) + \mathbf{A}_I^{1\mu}(x)\cdot\mathbf{T}_I + \mathbf{A}_I^{2\mu}(x)\cdot\mathbf{T}_I + B^{1\mu} + B^{2\mu}]\}\psi = 0 \qquad (3.4)$$

Under an $\Omega$-transformation, as defined in eq. 2.19 we use the <u>1</u> of SU(3) and <u>64</u> of U(64) to rotate the interaction fields $\mathbf{A}_I^{1\mu}(x)$ and $\mathbf{A}_I^{2\mu}(x)$, and the generators $\mathbf{T}_I$. We use the <u>3</u> of

[9] The periodic table consists of four layers. Each layer has 48 fermions consisting of 32 normal fermions in four generations, and 16 Dark fermions in four generations. See Blaha (2016a), (2016b) and (2016c) for details.

SU(3) and 64 of U(64) to rotate the $A_\Omega^{1\mu}(x)$ and $A_\Omega^{2\mu}(x)$ fields, and the fermion vector $\psi$. We then find the Dirac equation transforms to

$$\{\partial^\mu + i\,[g_\Omega A'_\Omega{}^{1\mu}(x) + g_\Omega A'_\Omega{}^{2\mu}(x) + \mathbf{A}'_I{}^{1\mu}(x)\cdot\mathbf{T}'_I + \mathbf{A}'_I{}^{2\mu}(x)\cdot\mathbf{T}'_I + B'^{1\mu} + B'^{2\mu}]\}\psi = 0 \qquad (3.5)$$

where

$$A'_\Omega{}^{1\mu}(x) = C_\Omega(x)A_\Omega{}^{1\mu}(x)C_\Omega{}^{-1}(x) - i\,C_\Omega(x)\partial^\mu C_\Omega{}^{-1}(x)/g_\Omega \qquad (3.6)$$
$$A'_\Omega{}^{2\mu}(x) = C_\Omega(x)A_\Omega{}^{2\mu}(x)C_\Omega{}^{-1}(x)$$
$$\mathbf{A}'_I{}^{1\mu}(x) = \mathbf{A}_I{}^{1\mu}(x)C^{-1}{}_\Omega(x)$$
$$\mathbf{A}'_I{}^{2\mu}(x) = \mathbf{A}_I{}^{2\mu}(x)C^{-1}{}_\Omega(x)$$
$$\mathbf{T}'_I = C_\Omega(x)\mathbf{T}_I$$
$$B'^{1\mu} = B^{1\mu}$$
$$B'^{2\mu} = B^{2\mu}$$

The effect of the Ω-transformation is to rotate the gauge fields. It is accompanied by a rotation of the generator matrices. Together they define an equivalent formulation of the original Dirac equation and thus the fermion sector. The next section provides an explicit simple example of Ω-transformations – an ElectroWeak theory.

Note that an Ω-transformation causes a change of gauge in the $A_\Omega{}^{1\mu}(x)$ field. The other gauge fields, $\mathbf{A}_I{}^{1\mu}(x)$ and $\mathbf{A}_I{}^{2\mu}(x)$, are 'rotated' but do not undergo a change of gauge. (Each of these other gauge fields do undergo their own particular changes of gauge for their transformation groups.)

The spinor connection field $B^{1\mu}$ (which Weinberg (1972) denotes as $\Gamma^\mu$ on p. 368) is not affected by Ω-transformations since it, as well as the General Coordinate affine connection, is in the gravitation sector, which all fermions (and bosons) experience uniformly.

## 3.4 Fermion Periodic Table and the 192 Dirac Spinor Vector ψ

The Periodic Table of 192 Fermions was constructed based on a number of reasonable (in the author's view) assumptions.[10] First we required that the fermion species (types) be constructed by complex Lorentz group transformations on a Dirac spinor at rest. This yielded four species of fermions that we identified with charged leptons, neutral leptons, up-type quarks, and down-type quarks. We then found that quarks occupied the 3 of color SU(3). Color SU(3) as well as the normal and Dark Weak interactions of group SU(2), and the normal and Dark electromagnetic interactions were 'derived' from the form of a Lorentz transformation.[11] Thus we found eight species of normal fermions, and four species of Dark fermions (since Dark fermions do not have color interactions – they are SU(3) singlets.) Thus in a generation there are eight normal fermions and four Dark fermions.

[10] The contents of this section and the figure appears in several of the author's earlier books in 2015 and 2016.
[11] These points all appear in Blaha (2015a).

Then noting that there are four conserved particle number operators (baryon number,[12] lepton number, Dark baryon number and Dark lepton number) we were led to posit a U(4) symmetry that generated four generations of fermions. Thus we found 32 normal fermions and 16 Dark fermions = 48 fermions in the four fenerations.

Next we noted that the number of particles in each of the four generations was (almost) conserved. This fact led to another group – the U(4) Layer group – that led to four layers of fermions. The total number of fermions then was found to be 4*48 = 192 fermions.

The Periodic Table that this construction implies appears in Fig. 3.1.

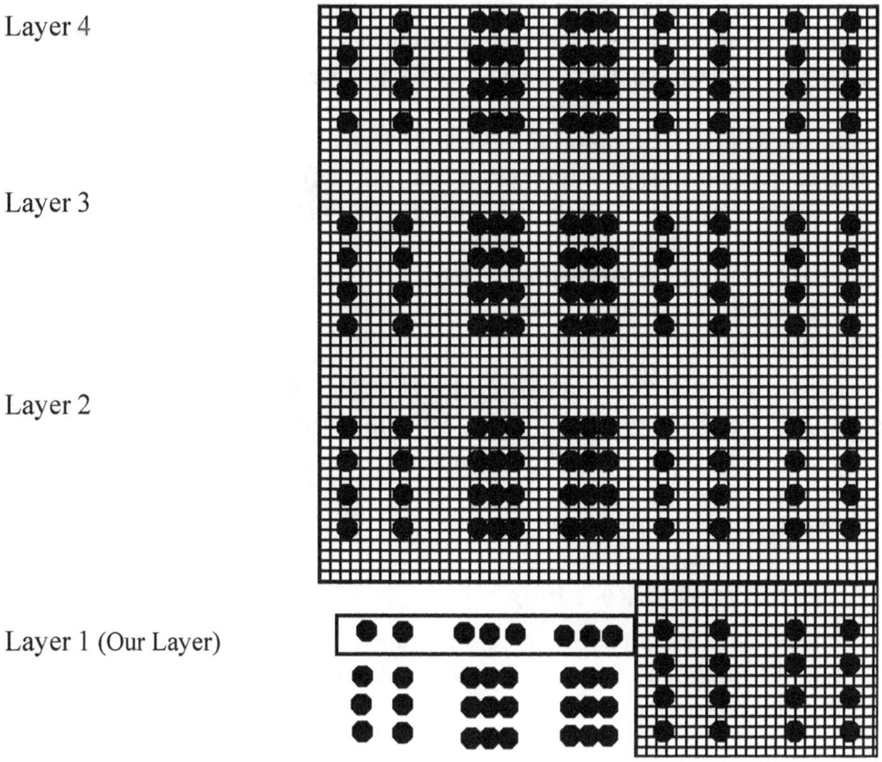


Figure 3.1. Dark parts of the periodic table are 'cross-hatched.' Light parts are the known fermions – with an additional, as yet not found, 4$^{th}$ generation of layer 1 is shown boxed. It is part of 'Dark matter' at present. When found experimentally it will be 'non-Dark.'

[12] The author has pointed out in previous books that disparities in measurements of the gravitational constant G could reflect the existence of a baryonic force that, like electromagnetism, leads to a conserved baryon number. Similar comments would apply to other conserved Generation group numbers and almost conserved Layer group numbers.

## 3.5 Structure of 192 Dirac Spinor Vector ψ

The Dirac spinor ψ for the 192 fermions is an eigenvector of SU3)⊗U(64). We can express ψ in the form ψ = $\{\psi_{ij}\}$ where i = 1, 2, 3 is an SU(3) index and j = 1, 2, … , 64 is a U(64) index. Thus for each value of j there is a triplet of spinor fields. Since there are 32 normal quark triplets in the periodic table totaling 96 quarks it is natural to identify these 32 triplets as also SU(3) triplets of the Ω interaction.[13] The remaining 96 fermions can be structured as 32 Ω triplets as well. We define these triplets as follows:

- In each generation of each layer we form an UP-type triplet consisting of two normal charged leptons plus one up-type Dark quark.
- In each generation of each layer we form an DOWN-type triplet consisting of two normal neutral leptons plus one down-type Dark quark.

Thus we have a structuring of the 192 fermion field vector into 64 triplets.

It is possible to reorder the structure of ψ and $A_\Omega{}^{1\mu}(x)$ and $A_\Omega{}^{2\mu}(x)$ with U(192) transformations. These transformations will also change the ordering of the eight gauge interaction generator matrices $\mathbf{T}_I$. A U(192) reordering does not have appear to have physical consequences – it can be viewed as a bookkeeping change.

## 3.6 Broken Ώ-Symmetry

If, on the contrary, we take Ώ-Symmetry to be U(192),[14] then it is broken at several levels to yield the eight interactions of which it is composed. Firstly, it is broken to SU3)⊗U(64), and then it is broken to the eight interactions. We anticipate that the breakdown takes place in one step through some form of Higgs Mechanism for masses and, perhaps, for coupling constants as we described in Blaha(2015d).

## 3.7 Example: ElectroWeak-like Theory

ElectroWeak model is an example of a global Ώ-transformation which does not include the full gamut of features outlined above. Focussing on the Weak and Electromagnetic interactions we consider the covariant derivative

$$\{\partial^\mu + i\,[g\mathbf{W}^\mu\cdot\boldsymbol{\tau} + g' W_0{}^\mu \tau_0]\}\psi = 0 \tag{3.7}$$

Rotating $W_3{}^\mu$ and $W_0{}^\mu$ with $C^{-1}{}_\Omega$

[13][13] Although we identify the triplets we still regard the interactions as separate and distinct.

[14] The choice of U(192) is dictated by the number of fundamental fermions, which, in turn, is ultimately dictated by space-time geometry.

$$\begin{bmatrix} gW_3^{\mu} \\ \\ g'W_0^{\mu} \end{bmatrix} \rightarrow \begin{bmatrix} g\cos\theta\ Z^{\mu} + g\sin\theta\ A^{\mu} \\ \\ -g'\sin\theta\ Z^{\mu} + g'\cos\theta\ A^{\mu} \end{bmatrix}$$

and rotating the generator matrices by $C_{\Omega}$

$$\begin{bmatrix} \tau_3 \\ \\ \tau_0 \end{bmatrix} \rightarrow \begin{bmatrix} \cos\theta\ \tau_3\ - \sin\theta\ \tau_0 \\ \\ \sin\theta\ \tau_3 + \cos\theta\ \tau_0 \end{bmatrix} \qquad (3.8)$$

and making an appropriate choice of $\theta_W$ yields the electromagnetic field A and Z

$$g'W_0^{\mu}\ \tau_0 + gW_3^{\mu}\tau_3 \rightarrow Z^{\mu}[(g\cos^2\theta - g'\sin^2\theta)\tau_3 - \tfrac{1}{2}\sin(2\theta)(g + g')\tau_0 + A^{\mu}[\tfrac{1}{2}\sin(2\theta)(g + g')\tau_3 + \\ + (g'\cos^2\theta - g\sin^2\theta)\tau_0] \qquad (3.9)$$

If we choose the coefficient of $A^{\mu}$ be e, then

$$e(\tau_3 + \tau_0) = [\tfrac{1}{2}\sin(2\theta)(g + g')\tau_3 + (g'\cos^2\theta - g\sin^2\theta)\tau_0] \qquad (3.10)$$

and thus

$$\tfrac{1}{2}\sin(2\theta)(g + g') = (g'\cos^2\theta - g\sin^2\theta) = e \qquad (3.11)$$

Consequently

$$g' = -(\tfrac{1}{2}\sin(2\theta) - \sin^2\theta)/(\tfrac{1}{2}\sin(2\theta) - \cos^2\theta) \qquad (3.12)$$

If we further choose g = g' (since equal coupling constants is an often stated goal of theorists) for simplicity we find the angle $\theta = \pi/8$ or 22.5° while the usual Weinberg angle $\theta_W$ is about 30°. Thus we find a close similarity to standard ElectroWeak theory.[15]

We conclude that ElectroWeak theory can be viewed as a subsector of the broken Ω-Symmetry theory.

[15] We can, of course, make the angles equal by adjusting the values of g ang g'.

## 3.8 Path Integral Formulation, and the Faddeev-Popov Method

The path integral formulation of our theory with the complete set of eleven integrations is fairly straightforward with one exception. Since we use complex valued coordinates for the color SU(3) gauge theory a somewhat different approach must be followed for it. We detail this approach in Blaha (2015a).

In this section we will explicitly consider the Faddeev-Popov Method for the Ω gauge field $A_\Omega{}^{1\mu}(x)$ only – noting that the Yang-Mills gauge fields of the other interactions must also be subject to gauge fixing using the Faddeev-Popov Method.

The path integral we consider for the $A_\Omega{}^{1\mu}(x)$ gauge field is:[16]

$$Z(J^\mu) = N\int DA_\Omega{}^{1\mu}\,\Delta(A_\Omega{}^1)\delta(F(A_\Omega))\exp\{i\int d^4y[\mathscr{L} + J^\mu(y)\,A_\Omega{}^1{}_\mu(y)]\} \tag{3.13}$$

where $\delta(F(A_\Omega))$ specifies the gauge, and $\Delta(A_\Omega{}^1)$ is its Faddeev-Popov determinant. The Faddeev-Popov determinant can be calculated in the standard way.[17]

First we consider the gauge fixing delta function. Note that it can be written as a delta function in the gauge times a determinant:

$$\delta(F(A_\Omega{}^\omega)) = \delta(\omega - \omega_0)|\det \delta F(A_\Omega{}^1{}_\mu{}^\omega(x))/\delta\omega(x)|^{-1}|_{F(A_\Omega)=0} \tag{3.14}$$

where $\omega_0$ is a reference gauge, where

$$\begin{aligned} A_\Omega{}^{1a}{}_\mu{}^\omega(x) &= A_\Omega{}^{1a}{}_\mu(x) - g_\Omega{}^{-1}\partial_\mu\omega^a + f_\Omega{}^{abc}\,\omega^b(x)A_\Omega{}^{1c}{}_\mu(x) \\ &= A_\Omega{}^{1a}{}_\mu(x) + \delta A_\Omega{}^{1a}{}_\mu{}^\omega(x) \end{aligned} \tag{3.15}$$

and where

$$F_A{}^a{}_\mu{}^\omega = F_A{}^a{}_\mu + f_\Omega{}^{abc}\,\omega^b(x)F_A{}^c{}_\mu \tag{3.16}$$

under an infinitesimal gauge transformation where the $f_\Omega{}^{abc}$ are SU3)⊗U(64) structure constants, and a, b, and c label the 72 generators of SU3)⊗U(64).

Also

$$\Delta(A_\Omega{}^1) = |\det \delta F(A_\Omega{}^1{}_\mu{}^\omega(x))/\delta\omega(x)||_{F(A_\Omega)=0,\,\omega=0} \tag{3.17}$$

We will choose the Lorentz gauge to evaluate the Faddeev-Popov determinant:

$$F^a(A_\Omega) = \partial_\mu A_\Omega{}^{1a\mu}(x) = 0 \tag{3.18}$$

We find

$$F^a(A_{\Omega\mu}{}^\omega(x)) = \partial^\mu[A_\Omega{}^{1a}{}_\mu(x) - g_\Omega{}^{-1}\partial_\mu\omega^a(x) + f_\Omega{}^{abc}\,\omega^b(x)A_\Omega{}^{1c}{}_\mu(x)]$$

---

[16] The $A_\Omega{}^{2\mu}(x)$ transforms homogeneously and thus does not have a Faddeev-Popov determinant $\Delta(A_\Omega{}^2)$.

[17] See for example Huang (1992).

$$= -g_\Omega^{-1}\partial^\mu\partial_\mu\omega^a(x) + f_\Omega^{abc}A_\Omega^{1c}{}_\mu(x)\,\partial^\mu\omega^b(x) \quad (3.19)$$

Thus

$$\delta F^a(A_{\Omega\mu}{}^\omega(x))/\delta\omega^b(x) = -g_\Omega^{-1}\delta^{ab}\partial^\mu\partial_\mu + f_\Omega^{abc}A_\Omega^{1c\mu}(x)\partial_\mu \quad (3.20)$$

and

$$\Delta(A_\Omega) = |\det(g_\Omega^{-1}\delta^{ab}\partial^\mu\partial_\mu - f_\Omega^{abc}A_\Omega^{1c\mu}(x)\partial_\mu)| \quad (3.21)$$

where | ... | represents the absolute value.

We can rewrite the Faddeev-Popov determinant as a path integral over anti-commuting c-number fields $\chi^a$ with a ghost Lagrangian:

$$\Delta(A_\Omega) = \int D\chi^* D\chi \exp[\,i\int d^4x\, \mathscr{L}_\Omega^{ghost}(x)] \quad (3.22)$$

where

$$\mathscr{L}_\Omega{}^{ghost}(x) = \chi^{a*}(x)[\delta^{ab}\partial^\mu\partial_\mu - gf_\Omega^{abc}A_\Omega^{1c\mu}(x)\partial_\mu]\chi^b(x) \quad (3.23)$$

with a, b and c ranging from 1 through 192.

The Ω gauge field lagrangian thus acquires a ghost lagrangian:

$$Z(J^\mu) = N\int DA_\Omega^{1\mu}\,\delta(F(A_\Omega))\exp\{i\int d^4y[\mathscr{L}_\Omega{}^{ghost}(y) + J^\mu(y)\,A_\Omega{}^1{}_\mu(y)]\} \quad 3.24)$$

in addition to other factors for the other gauge field interactions.

## 3.9 Interactions of $A_\Omega{}^\mu(x)$ and Ghost 'Fields'

In general the $A_\Omega{}^\mu(x)$ field with its 192 components embodies interactions between all 192 fermions in the Fermion Periodic Table. Similarly, the 192 ghost fields also yield interactions between all 192 fermions through their affect on $A_\Omega{}^\mu$ quanta propagated between frmions.

The $A_\Omega{}^\mu(x)$ field quanta has in and out states in perturbation theory. Ghost fields only exist as interactions between $A_\Omega{}^\mu$ quanta within Feynman diagrams and do not have in or out states. In this section we will overview the $A_\Omega{}^\mu(x)$ and ghost interactions.

### 3.9.1 $A_\Omega{}^\mu(x)$ Interactions

$A_\Omega{}^\mu$ quanta have self-interactions that are qualitatively similar to those of other Yang-Mills gauge fields. They are 'more numerous' because the $A_\Omega{}^\mu$ field is an SU3)⊗U(64) gauge field.

$A_\Omega{}^\mu$ quanta can be exchanged between any pair of fermions. As a result quarks and leptons interact via $A_\Omega{}^\mu$ quanta exchange (with possibly $A_\Omega{}^\mu$ self-interactions within the quanta exchanges) Since the only known interactions between quarks and leptons are Weak, and Electromagnetic interactions, the coupling constant $g_\Omega$ must be very small and/or the mass of the $A_\Omega{}^\mu$ quanta is very large. Since the Ω-symmetry is clearly broken, it is likely that $A_\Omega{}^\mu$ quanta acquire a mass via the Higgs Mechanism.

## 3.10 A New Form of Unification

*Perhaps the most remarkable property of the SU3)⊗U(64) Ω-symmetry is its universality in coupling to all fermions. This unifying feature of Ω-symmetry is a new Unification Paradigm beyond the three possible forms of unification specified in I. It enables us to view the origin of the universe – the Big Bang – as unified about Ω-symmetry – not as the 'separate' appearance of 192 different fermions and 11 different interactions. The Ω-symmetry, the 11$^{th}$ interaction, ties the Big Bang universe together.*

*Indeed we can view the various fermions as being generated from a primordial sea of* $A_\Omega{}^\mu$ *quanta that 'decay' into pairs of different fermions of the set of 192 fermions. We note that all fermions (and bosons) may be expected to be massless at the Big Bang instant.*

## 3.11 Ghost Field Interactions

Ghost fields interact with $A_\Omega{}^\mu$ quanta in a manner analogous to ghost field interactions in other gauge theories. They do not interact directly with fermions – but rather modify the $A_\Omega{}^\mu$ interaction between fermions by changing the $A_\Omega{}^\mu$ propagator.

Ghost fields do not have in-states or out-states in perturbation theory. Thus they are not part of states contributing to unitarity sums.

# 4. The Total Covariant Derivative

We begin by defining a 'space' vector which also is a fundamental representation vector of SU(3)⊗SU(2)⊗U(1)⊗SU(2)⊗U(1)⊗U(4)⊗U(4)⊗U(4)⊗SU(3)⊗U(64).[18]

$$V_\sigma = V_\sigma^{\,aijkmno}(x)\gamma_a T_i \tau_j \tau_{Dj} \mathbf{G}_{U(4)k} G_{LU(4)m} \mathbf{G}_{RflatU(4)n} G_{SU(3)\otimes U(64)o} \tag{4.1}$$

where $\boldsymbol{\tau}_D$ represents the Dark SU(2) generators and $G_{SU(3)\otimes U(64)}$ is the set of SU(3)⊗U(64) generators.

Then we use the following generalized covariant derivative of this vector: [19,20]

$$\begin{aligned} D_\nu V_\mu &= (\partial_\nu + iF_\nu)V_\mu - H^\sigma{}_{\nu\mu} V_\sigma \\ &= [g^\sigma{}_\mu \partial_\nu + ig^\sigma{}_\mu F_\nu - H^\sigma{}_{\nu\mu}]V_\sigma \\ &= [g^\sigma{}_\mu \partial_\nu + iD^\sigma{}_{\mu\nu}]V_\sigma \end{aligned} \tag{4.2}$$

where[21]

$$F^\mu = g_\Omega A_\Omega^{\,1\mu}(x) + g_\Omega A_\Omega^{\,2\mu}(x) + \mathbf{A}_I^{\,1\mu}(x)\cdot\mathbf{T}_I + \mathbf{A}_I^{\,2\mu}(x)\cdot\mathbf{T}_I + B^{1\mu} + B^{2\mu} \tag{4.3}$$

by eq. 3.4, and

$$H^\sigma{}_{\nu\mu} = \Gamma_{GR}{}^\sigma{}_{\nu\mu} + \Gamma_{GR}{}^{2\sigma}{}_{\nu\mu} \tag{4.4}$$

$$D^\sigma{}_{\mu\nu} = g^\sigma{}_\mu F_\nu + iH^\sigma{}_{\nu\mu} \tag{4.5}$$

where we have abstracted the complex part of the complex affine connection into the U(4) gauge field. See I for a detailed discussion. Eq. 4.4 is the real-valued part of the affine connection.

*Commutators of the vector fields in $F_\mu$ are implicit when the covariant derivative is applied to vectors and tensors such as $V_\sigma$.*

---

[18] Much of the material in this chapter appeared in Blaha (2016h) and earlier books.

[19] We use the superscript '1' to distinguish primary connections from secondary connections labeled '2'. The discussion in this subsection and in the following subsection parallels that of chapter 4 of I.

[20] Commutator 'cross products' are usually implicit in the following equations in chapter 5.

[21] We will omit the insertion of coupling constants of $B^{1\mu}$ and $B^{2\mu}$ in the interests of simplifying the expressions.

# 5. The Riemann-Christoffel Tensor

Eqs. 4.1 – 4.5 enables us to calculate the Riemann-Christoffel curvature tensor, and then its contractions $R_{\mu\nu}$ and R using[22]

$$(D_\nu D_\mu - D_\mu D_\nu)V_\sigma = R^\beta{}_{\sigma\nu\mu}V_\beta \tag{5.1}$$

The second order covariant derivative of $V_\sigma$ is

$$D_\nu D_\mu V_\sigma = \{g^\alpha{}_\mu(\partial_\nu + iF_\nu) - H^\alpha{}_{\mu\nu}\}\{g^\beta{}_\sigma(\partial_\alpha + iF_\alpha)V_\beta - H^\beta{}_{\sigma\alpha}V_\beta\} - H^\gamma{}_{\nu\sigma}\{g^\alpha{}_\gamma(\partial_\mu + iF_\mu)V_\alpha - H^\alpha{}_{\gamma\mu}V_\alpha\} \tag{5.2}$$

with implicit commutators with the guage field terms.

## 5.1 The Eleven Interaction Riemann-Christoffel Curvature Tensor

Using the definitions in chapter 4 we find

$$\begin{aligned}
R'^\beta{}_{\sigma\nu\mu}V_\beta = {} & g^\alpha{}_\mu(\partial_\nu + iF_\nu)g^\beta{}_\sigma(\partial_\alpha + iF_\alpha)V_\beta - H^\alpha{}_{\mu\nu}g^\beta{}_\sigma(\partial_\alpha + iF_\alpha)V_\beta + \\
& + H^\alpha{}_{\mu\nu}H^\beta{}_{\sigma\alpha}V_\beta - g^\alpha{}_\mu(\partial_\nu + iF_\nu)H^\beta{}_{\sigma\alpha}V_\beta - H^\gamma{}_{\nu\sigma}\{g^\alpha{}_\gamma(\partial_\mu + iF_\mu)V_\alpha - H^\alpha{}_{\gamma\mu}V_\alpha\} - \\
& - \{\mu\leftrightarrow\nu\}
\end{aligned}$$

$$= ig^\beta{}_\sigma(\partial_\nu F_\mu - \partial_\mu F_\nu - i[F_\nu, F_\mu])V_\beta + (\partial_\mu H^\beta{}_{\sigma\nu} - \partial_\nu H^\beta{}_{\sigma\mu} + H^\gamma{}_{\nu\sigma}H^\beta{}_{\gamma\mu} - H^\gamma{}_{\mu\sigma}H^\beta{}_{\gamma\nu})V_\beta$$

$$\begin{aligned}
= {} & ig^\beta{}_\sigma(F_E{}^1{}_{\nu\mu} + F_E{}^2{}_{\nu\mu} + F_W{}^1{}_{\nu\mu} + F_W{}^2{}_{\nu\mu} + F_{DE}{}^1{}_{\nu\mu} + F_{DE}{}^2{}_{\nu\mu} + F_{DW}{}^1{}_{\nu\mu} + F_{DW}{}^2{}_{\nu\mu} + F_{SU(3)}{}^1{}_{\nu\mu} + \\
& + F_{SU(3)}{}^2{}_{\nu\mu} + F_U{}^1{}_{\nu\mu} + F_U{}^2{}_{\nu\mu} + F_V{}^1{}_{\nu\mu} + F_V{}^2{}_{\nu\mu} + F_\Omega{}^1{}_{\nu\mu} + F_\Omega{}^2{}_{\nu\mu})V_\beta + \\
& + (ig^\beta{}_\sigma B^1{}_{\nu\mu} + ig^\beta{}_\sigma B^2{}_{\nu\mu} + \partial_\mu H^\beta{}_{\sigma\nu} - \partial_\nu H^\beta{}_{\sigma\mu} + H^\gamma{}_{\nu\sigma}H^\beta{}_{\gamma\mu} - H^\gamma{}_{\mu\sigma}H^\beta{}_{\gamma\nu})V_\beta
\end{aligned}$$

$$\begin{aligned}
= {} & R'_E{}^\beta{}_{\sigma\nu\mu}V_\beta + R'_{SU(2)}{}^\beta{}_{\sigma\nu\mu}V_\beta + R'_{DE}{}^\beta{}_{\sigma\nu\mu}V_\beta + R'_{DSU(2)}{}^\beta{}_{\sigma\nu\mu}V_\beta + R'_{SU(3)}{}^\beta{}_{\sigma\nu\mu}V_\beta + R'_U{}^\beta{}_{\sigma\nu\mu}V_\beta + \\
& + R'_V{}^\beta{}_{\sigma\nu\mu}V_\beta + R'_{Rflat}{}^\beta{}_{\sigma\nu}V_\beta + R'_\Omega{}^\beta{}_{\sigma\nu}V_\beta + R'_B{}^\beta{}_{\sigma\nu}V_\beta + R'_G{}^\beta{}_{\sigma\nu\mu}V_\beta
\end{aligned} \tag{5.3}$$

where

$$\begin{aligned}
R'_{SU(3)}{}^\beta{}_{\sigma\nu\mu} &= ig^\beta{}_\sigma(F_{SU(3)}{}^1{}_{\nu\mu} + F_{SU(3)}{}^2{}_{\nu\mu}) \\
R'_{SU(2)}{}^\beta{}_{\sigma\nu\mu} &= ig^\beta{}_\sigma(F_W{}^1{}_{\nu\mu} + F_W{}^2{}_{\nu\mu}) \\
R'_E{}^\beta{}_{\sigma\nu\mu} &= ig^\beta{}_\sigma(F_E{}^1{}_{\nu\mu} + F_E{}^2{}_{\nu\mu}) \\
R'_U{}^\beta{}_{\sigma\nu\mu} &= ig^\beta{}_\sigma(F_U{}^1{}_{\nu\mu} + F_U{}^2{}_{\nu\mu}) \\
R'_V{}^\beta{}_{\sigma\nu\mu} &= ig^\beta{}_\sigma(F_V{}^1{}_{\nu\mu} + F_V{}^2{}_{\nu\mu})
\end{aligned} \tag{5.4}$$

---

[22] Much of the material in this chapter appeared in Blaha (2016h) and earlier books.

$$
\begin{aligned}
R'_{DSU(2)}{}^{\beta}{}_{\sigma\nu\mu} &= ig^{\beta}{}_{\sigma}(F_{DW}{}^{1}{}_{\nu\mu} + F_{DW}{}^{2}{}_{\nu\mu}) \\
R'_{DE}{}^{\beta}{}_{\sigma\nu\mu} &= ig^{\beta}{}_{\sigma}(F_{DE}{}^{1}{}_{\nu\mu} + F_{DE}{}^{2}{}_{\nu\mu}) \\
R'_{Rflat}{}^{\beta}{}_{\sigma\nu\mu} &= ig^{\beta}{}_{\sigma}(F_{Rflat}{}^{1}{}_{\nu\mu} + F_{Rflat}{}^{2}{}_{\nu\mu}) \\
R'_{\Omega}{}^{\beta}{}_{\sigma\nu\mu} &= ig^{\beta}{}_{\sigma}(F_{\Omega}{}^{1}{}_{\nu\mu} + F_{\Omega}{}^{2}{}_{\nu\mu}) \\
R'_{B}{}^{\beta}{}_{\sigma\nu\mu} &= ig^{\beta}{}_{\sigma}(F_{B}{}^{1}{}_{\nu\mu} + F_{B}{}^{2}{}_{\nu\mu})
\end{aligned}
$$

and

$$
\begin{aligned}
R'_{G}{}^{\beta}{}_{\sigma\nu\mu} = {} & \partial_\mu H^{1\beta}{}_{\sigma\nu} - \partial_\nu H^{1\beta}{}_{\sigma\mu} + H^{1\gamma}{}_{\nu\sigma}H^{1\beta}{}_{\gamma\mu} - H^{1\gamma}{}_{\mu\sigma}H^{1\beta}{}_{\gamma\nu} + \partial_\mu H^{2\beta}{}_{\sigma\nu} - \partial_\nu H^{2\beta}{}_{\sigma\mu} + \\
& + H^{2\gamma}{}_{\nu\sigma}H^{2\beta}{}_{\gamma\mu} - H^{2\gamma}{}_{\mu\sigma}H^{2\beta}{}_{\gamma\nu} + H^{1\gamma}{}_{\nu\sigma}H^{2\beta}{}_{\gamma\mu} - H^{1\gamma}{}_{\mu\sigma}H^{2\beta}{}_{\gamma\nu} + H^{2\gamma}{}_{\nu\sigma}H^{1\beta}{}_{\gamma\mu} - \Gamma^{2\gamma}{}_{\mu\sigma}\Gamma^{\beta}{}_{\gamma\nu} \qquad (5.5) \\
& = R^{1\beta}{}_{\sigma\nu\mu} + R^{2\beta}{}_{\sigma\nu\mu}
\end{aligned}
$$

with

$$
\begin{aligned}
H^{\beta}{}_{\sigma\nu\mu} &= \partial_\mu H^{\beta}{}_{\sigma\nu} - \partial_\nu H^{\beta}{}_{\sigma\mu} + H^{\gamma}{}_{\nu\sigma}H^{\beta}{}_{\gamma\mu} - H^{\gamma}{}_{\mu\sigma}H^{\beta}{}_{\gamma\nu} \qquad (5.6) \\
R^{1\beta}{}_{\sigma\nu\mu} &= \partial_\mu H^{1\beta}{}_{\sigma\nu} - \partial_\nu H^{1\beta}{}_{\sigma\mu} + H^{1\gamma}{}_{\nu\sigma}H^{1\beta}{}_{\gamma\mu} - H^{1\gamma}{}_{\mu\sigma}H^{1\beta}{}_{\gamma\nu} \qquad (5.7) \\
R^{2\beta}{}_{\sigma\nu\mu\rho} &= \partial_\mu H^{2\beta}{}_{\sigma\nu} - \partial_\nu H^{2\beta}{}_{\sigma\mu} + H^{2\gamma}{}_{\nu\sigma}H^{2\beta}{}_{\gamma\mu} - H^{2\gamma}{}_{\mu\sigma}H^{2\beta}{}_{\gamma\nu} + \\
& \quad + H^{1\gamma}{}_{\nu\sigma}H^{2\beta}{}_{\gamma\mu} - H^{1\gamma}{}_{\mu\sigma}H^{2\beta}{}_{\gamma\nu} + H^{2\gamma}{}_{\nu\sigma}H^{1\beta}{}_{\gamma\mu} - H^{2\gamma}{}_{\mu\sigma}H^{1\beta}{}_{\gamma\nu} \qquad (5.8)
\end{aligned}
$$

and

$$
\begin{aligned}
H^{1\sigma}{}_{\nu\mu} &= \Gamma_{GR}{}^{\sigma}{}_{\nu\mu} \\
H^{2\sigma}{}_{\nu\mu} &= \Gamma_{GR}{}^{2\sigma}{}_{\nu\mu}
\end{aligned}
$$

and where

$$
\begin{aligned}
F_{SU(3)}{}^{1}{}_{\kappa\mu} &= \partial A_{SU(3)}{}^{1}{}_{\mu}/\partial x^{\kappa} - \partial A_{SU(3)}{}^{1}{}_{\kappa}/\partial x^{\mu} + ig_1[A_{SU(3)}{}^{1}{}_{\kappa}, A_{U(3)}{}^{1}{}_{\mu}] \qquad (5.9) \\
F_{W}{}^{1}{}_{\kappa\mu} &= \partial W^{1}{}_{\mu}/\partial x^{\kappa} - \partial W^{1}{}_{\kappa}/\partial x^{\mu} + ig_2[W^{1}{}_{\kappa}, W^{1}{}_{\mu}] \\
F_{E}{}^{1}{}_{\kappa\mu} &= \partial A_{E}{}^{1}{}_{\mu}/\partial x^{\kappa} - \partial A_{E}{}^{1}{}_{\kappa}/\partial x^{\mu} \\
F_{DW}{}^{1}{}_{\kappa\mu} &= \partial W_{D}{}^{1}{}_{\mu}/\partial x^{\kappa} - \partial W_{D}{}^{1}{}_{\kappa}/\partial x^{\mu} + ig_4[W_{D}{}^{1}{}_{\kappa}, W_{D}{}^{1}{}_{\mu}] \\
F_{DE}{}^{1}{}_{\kappa\mu} &= \partial A_{DE}{}^{1}{}_{\mu}/\partial x^{\kappa} - \partial A_{DE}{}^{1}{}_{\kappa}/\partial x^{\mu} \\
F_{U}{}^{1}{}_{\kappa\mu} &= \partial U^{1}{}_{\mu}/\partial x^{\kappa} - \partial U^{1}{}_{\kappa}/\partial x^{\mu} + ig_6[U^{1}{}_{\kappa}, U^{1}{}_{\mu}] \\
F_{V}{}^{1}{}_{\kappa\mu} &= \partial V^{1}{}_{\mu}/\partial x^{\kappa} - \partial V^{1}{}_{\kappa}/\partial x^{\mu} + ig_7[V^{1}{}_{\kappa}, V^{1}{}_{\mu}] \\
F_{Rflat}{}^{1}{}_{\kappa\mu} &= \partial A_{Rflat}{}^{1}{}_{\mu}/\partial x^{\kappa} - \partial A_{Rflat}{}^{1}{}_{\kappa}/\partial x^{\mu} + ig_8[A_{Rflat}{}^{1}{}_{\kappa}, A_{Rflat}{}^{1}{}_{\mu}] \\
F_{\Omega}{}^{1}{}_{\kappa\mu} &= \partial A_{\Omega}{}^{1}{}_{\mu}/\partial x^{\kappa} - \partial A_{\Omega}{}^{1}{}_{\kappa}/\partial x^{\mu} + ig_{\Omega}[A_{\Omega}{}^{1}{}_{\kappa}, A_{\Omega}{}^{1}{}_{\mu}] \\
F_{B}{}^{1}{}_{\kappa\mu} &= \partial B^{1}{}_{\mu}/\partial x^{\kappa} - \partial B^{1}{}_{\kappa}/\partial x^{\mu} + i[B^{1}{}_{\kappa}, B^{1}{}_{\mu}]
\end{aligned}
$$

$$
\begin{aligned}
F_{SU(3)}{}^{2}{}_{\kappa\mu} &= \partial A_{SU(3)}{}^{2}{}_{\mu}/\partial x^{\kappa} - \partial A_{SU(3)}{}^{2}{}_{\kappa}/\partial x^{\mu} + ig_1[A_{SU(3)}{}^{2}{}_{\kappa}, A_{SU(3)}{}^{2}{}_{\mu}] + ig_1[A_{SU(3)}{}^{1}{}_{\kappa}, A_{SU(3)}{}^{2}{}_{\mu}] + \\
& \quad + ig_1[A_{SU(3)}{}^{2}{}_{\kappa}, A_{SU(3)}{}^{1}{}_{\mu}] \\
F_{W}{}^{2}{}_{\kappa\mu} &= \partial W^{2}{}_{\mu}/\partial x^{\kappa} - \partial W^{2}{}_{\kappa}/\partial x^{\mu} + ig_2[W^{2}{}_{\kappa}, W^{2}{}_{\mu}] + ig_2[W^{1}{}_{\kappa}, W^{2}{}_{\mu}] + ig_2[W^{2}{}_{\kappa}, W^{1}{}_{\mu}] \\
F_{E}{}^{2}{}_{\kappa\mu} &= \partial A_{E}{}^{2}{}_{\mu}/\partial x^{\kappa} - \partial A_{E}{}^{2}{}_{\kappa}/\partial x^{\mu} \\
F_{DW}{}^{2}{}_{\kappa\mu} &= \partial W_{D}{}^{2}{}_{\mu}/\partial x^{\kappa} - \partial W_{D}{}^{2}{}_{\kappa}/\partial x^{\mu} + ig_4[W_{D}{}^{2}{}_{\kappa}, W_{D}{}^{2}{}_{\mu}] + ig_4[W_{D}{}^{1}{}_{\kappa}, W_{D}{}^{2}{}_{\mu}] + ig_4[W_{D}{}^{2}{}_{\kappa}, W_{D}{}^{1}{}_{\mu}] \\
F_{DE}{}^{2}{}_{\kappa\mu} &= \partial A_{DE}{}^{2}{}_{\mu}/\partial x^{\kappa} - \partial A_{DE}{}^{2}{}_{\kappa}/\partial x^{\mu} \\
F_{U}{}^{2}{}_{\kappa\mu} &= \partial U^{2}{}_{\mu}/\partial x^{\kappa} - \partial U^{2}{}_{\kappa}/\partial x^{\mu} + ig_6[U^{2}{}_{\kappa}, U^{2}{}_{\mu}] + ig_6[U^{1}{}_{\kappa}, U^{2}{}_{\mu}] + ig_6[U^{2}{}_{\kappa}, U^{1}{}_{\mu}]
\end{aligned}
$$

$$F_{V}{}^{2}{}_{\kappa\mu} = \partial V^{2}{}_{\mu}/\partial x^{\kappa} - \partial V^{2}{}_{\kappa}/\partial x^{\mu} + ig_{7}[V^{2}{}_{\kappa}, V^{2}{}_{\mu}] + ig_{7}[V^{1}{}_{\kappa}, V^{2}{}_{\mu}] + ig_{7}[V^{2}{}_{\kappa}, V^{1}{}_{\mu}]$$
$$F_{Rflat}{}^{2}{}_{\kappa\mu} = \partial A_{Rflat}{}^{2}{}_{\mu}/\partial x^{\kappa} - \partial A_{Rflat}{}^{2}{}_{\kappa}/\partial x^{\mu} + ig_{8}[A_{Rflat}{}^{2}{}_{\kappa}, A_{Rflat}{}^{2}{}_{\mu}] + ig_{8}[A_{Rflat}{}^{1}{}_{\kappa}, A_{Rflat}{}^{2}{}_{\mu}] + + ig_{8}[A_{Rflat}{}^{2}{}_{\kappa}, A_{Rflat}{}^{1}{}_{\mu}]$$
$$F_{\Omega}{}^{2}{}_{\kappa\mu} = \partial A_{\Omega}{}^{2}{}_{\mu}/\partial x^{\kappa} - \partial A_{\Omega}{}^{2}{}_{\kappa}/\partial x^{\mu} + ig_{\Omega}[A_{\Omega}{}^{2}{}_{\kappa}, A_{\Omega}{}^{2}{}_{\mu}] + ig_{\Omega}[A_{\Omega}{}^{1}{}_{\kappa}, A_{\Omega}{}^{2}{}_{\mu}] + ig_{\Omega}[A_{\Omega}{}^{2}{}_{\kappa}, A_{\Omega}{}^{1}{}_{\mu}]$$
$$F_{B}{}^{2}{}_{\kappa\mu} = \partial B^{2}{}_{\mu}/\partial x^{\kappa} - \partial B^{2}{}_{\kappa}/\partial x^{\mu} + i[B^{2}{}_{\mu}, B^{2}{}_{\kappa}] + i[B^{1}{}_{\mu}, B^{2}{}_{\kappa}] + i[B^{2}{}_{\mu}, B^{1}{}_{\kappa}]$$

Note that $R'^{\beta}{}_{\sigma\nu\mu}$ factorizes into U(1)⊗SU(2)⊗U(1)⊗SU(2)⊗SU(3)⊗U(4)⊗U(4)⊗U(4)⊗U(4)⊗U(64) parts and a Riemann-Christoffel Gravitational curvature tensor part. For later use in defining a lagrangian we define

$$R'^{\beta}{}_{\sigma\nu\mu} = R'_{E}{}^{1\beta}{}_{\sigma\nu\mu} + R'_{E}{}^{2\beta}{}_{\sigma\nu\mu} + R'_{SU(2)}{}^{1\beta}{}_{\sigma\nu\mu} + R'_{SU(2)}{}^{2\beta}{}_{\sigma\nu\mu} + R'_{DE}{}^{1\beta}{}_{\sigma\nu\mu} + R'_{DE}{}^{2\beta}{}_{\sigma\nu\mu} + R'_{DSU(2)}{}^{1\beta}{}_{\sigma\nu\mu} + + R'_{DSU(2)}{}^{2\beta}{}_{\sigma\nu\mu} + R'_{SU(3)}{}^{1\beta}{}_{\sigma\nu\mu} + R'_{SU(3)}{}^{2\beta}{}_{\sigma\nu\mu} + R'_{U}{}^{1\beta}{}_{\sigma\nu\mu} + R'_{U}{}^{2\beta}{}_{\sigma\nu\mu} + R'_{V}{}^{1\beta}{}_{\sigma\nu\mu} + R'_{V}{}^{2\beta}{}_{\sigma\nu\mu} + + R'_{Rflat}{}^{1\beta}{}_{\sigma\nu\mu} + R'_{Rflat}{}^{2\beta}{}_{\sigma\nu\mu} + R'_{\Omega}{}^{1\beta}{}_{\sigma\nu\mu} + R'_{\Omega}{}^{2\beta}{}_{\sigma\nu\mu} + R'_{B}{}^{1\beta}{}_{\sigma\nu\mu} + R'_{B}{}^{2\beta}{}_{\sigma\nu\mu} + R^{1\beta}{}_{\sigma\nu\mu} + R^{2\beta}{}_{\sigma\nu\mu} \qquad (5.10)$$

where

$$R'_{E}{}^{1\beta}{}_{\sigma\nu\mu} = ig^{\beta}{}_{\sigma}F_{E}{}^{1}{}_{\nu\mu}$$
$$R'_{E}{}^{2\beta}{}_{\sigma\nu\mu} = ig^{\beta}{}_{\sigma}F_{DE}{}^{2}{}_{\nu\mu}$$

$$R'_{DE}{}^{1\beta}{}_{\sigma\nu\mu} = ig^{\beta}{}_{\sigma}F_{E}{}^{1}{}_{\nu\mu}$$
$$R'_{DE}{}^{2\beta}{}_{\sigma\nu\mu} = ig^{\beta}{}_{\sigma}F_{DE}{}^{2}{}_{\nu\mu}$$

$$R'_{SU(2)}{}^{1\beta}{}_{\sigma\nu\mu} = ig^{\beta}{}_{\sigma}F_{W}{}^{1}{}_{\nu\mu}$$
$$R'_{SU(2)}{}^{2\beta}{}_{\sigma\nu\mu} = ig^{\beta}{}_{\sigma}F_{DW}{}^{2}{}_{\nu\mu}$$

$$R'_{DSU(2)}{}^{1\beta}{}_{\sigma\nu\mu} = ig^{\beta}{}_{\sigma}F_{W}{}^{1}{}_{\nu\mu}$$
$$R'_{DSU(2)}{}^{2\beta}{}_{\sigma\nu\mu} = ig^{\beta}{}_{\sigma}F_{DW}{}^{2}{}_{\nu\mu}$$

$$R'_{SU(3)}{}^{1\beta}{}_{\sigma\nu\mu} = ig^{\beta}{}_{\sigma}F_{SU(3)}{}^{1}{}_{\nu\mu}$$
$$R'_{SU(3)}{}^{2\beta}{}_{\sigma\nu\mu} = ig^{\beta}{}_{\sigma}F_{SU(3)}{}^{2}{}_{\nu\mu}$$

$$R'_{U}{}^{1\beta}{}_{\sigma\nu\mu} = ig^{\beta}{}_{\sigma}F_{U}{}^{1}{}_{\nu\mu}$$
$$R'_{U}{}^{2\beta}{}_{\sigma\nu\mu} = ig^{\beta}{}_{\sigma}F_{U}{}^{2}{}_{\nu\mu}$$

$$R'_{V}{}^{1\beta}{}_{\sigma\nu\mu} = ig^{\beta}{}_{\sigma}F_{V}{}^{1}{}_{\nu\mu}$$
$$R'_{V}{}^{2\beta}{}_{\sigma\nu\mu} = ig^{\beta}{}_{\sigma}F_{V}{}^{2}{}_{\nu\mu}$$

$$R'_{Rflat}{}^{1\beta}{}_{\sigma\nu\mu} = ig^{\beta}{}_{\sigma}F_{Rflat}{}^{1}{}_{\nu\mu}$$
$$R'_{Rflat}{}^{2\beta}{}_{\sigma\nu\mu} = ig^{\beta}{}_{\sigma}F_{Rflat}{}^{2}{}_{\nu\mu}$$

$$R'_{\Omega}{}^{1\beta}{}_{\sigma\nu\mu} = ig^{\beta}{}_{\sigma}F_{\Omega}{}^{1}{}_{\nu\mu}$$

$$R'_{\Omega}{}^{2\beta}{}_{\sigma\nu\mu} = ig^{\beta}{}_{\sigma}F_{\Omega}{}^{2}{}_{\nu\mu}$$

$$R'_{B}{}^{1\beta}{}_{\sigma\nu\mu} = ig^{\beta}{}_{\sigma}B^{1}{}_{\nu\mu}$$
$$R'_{B}{}^{2\beta}{}_{\sigma\nu\mu} = ig^{\beta}{}_{\sigma}B^{2}{}_{\nu\mu}$$

The total Ricci tensor is

$$R'_{\sigma\mu} = R'^{\beta}{}_{\sigma\beta\mu} \tag{5.10}$$

$$= iF_{E}{}^{1}{}_{\sigma\mu} + iF_{E}{}^{2}{}_{\sigma\mu} + iF_{W}{}^{1}{}_{\sigma\mu} + iF_{W}{}^{2}{}_{\sigma\mu} + iF_{DE}{}^{1}{}_{\sigma\mu} + iF_{DE}{}^{2}{}_{\sigma\mu} + iF_{DW}{}^{1}{}_{\sigma\mu} + iF_{DW}{}^{2}{}_{\sigma\mu} + iF_{SU(3)}{}^{1}{}_{\sigma\mu} + iF_{SU(3)}{}^{2}{}_{\sigma\mu} +$$
$$+ iF_{U}{}^{1}{}_{\sigma\mu} + iF_{U}{}^{2}{}_{\sigma\mu} + iF_{V}{}^{1}{}_{\sigma\mu} + iF_{V}{}^{2}{}_{\sigma\mu} + iF_{Rflat}{}^{1}{}_{\sigma\mu} + iF_{Rflat}{}^{2}{}_{\sigma\mu} + iF_{\Omega}{}^{1}{}_{\sigma\mu} + iF_{\Omega}{}^{2}{}_{\sigma\mu} + iB^{1}{}_{\sigma\mu} + iB^{2}{}_{\sigma\mu} +$$
$$+ \partial_{\mu}H^{1\beta}{}_{\sigma\beta} - \partial_{\beta}H^{1\beta}{}_{\sigma\mu} + H^{1\gamma}{}_{\beta\sigma}H^{1\beta}{}_{\gamma\mu} - H^{1\gamma}{}_{\mu\sigma}H^{1\beta}{}_{\gamma\beta} +$$
$$+ \partial_{\mu}H^{2\beta}{}_{\sigma\beta} - \partial_{\beta}H^{2\beta}{}_{\sigma\mu} + H^{2\gamma}{}_{\beta\sigma}H^{2\beta}{}_{\gamma\mu} - H^{2\gamma}{}_{\mu\sigma}H^{2\beta}{}_{\gamma\beta} + H^{1\gamma}{}_{\beta\sigma}H^{2\beta}{}_{\gamma\mu} - H^{1\gamma}{}_{\mu\sigma}H^{2\beta}{}_{\gamma\beta} + H^{2\gamma}{}_{\beta\sigma}H^{1\beta}{}_{\gamma\mu} - H^{2\gamma}{}_{\mu\sigma}H^{1\beta}{}_{\gamma\beta}$$

$$= R'_{E}{}^{1}{}_{\sigma\mu} + R'_{E}{}^{2}{}_{\sigma\mu} + R'_{SU(2)}{}^{1}{}_{\sigma\mu} + R'_{SU(2)}{}^{2}{}_{\sigma\mu} + R'_{DE}{}^{1}{}_{\sigma\mu} + R'_{DE}{}^{2}{}_{\sigma\mu} + R'_{DSU(2)}{}^{1}{}_{\sigma\mu} + R'_{DSU(2)}{}^{2}{}_{\sigma\mu} + R'_{SU(3)}{}^{1}{}_{\sigma\mu} +$$
$$+ R'_{SU(3)}{}^{2}{}_{\sigma\mu} + R'_{U}{}^{1}{}_{\sigma\mu} + R'_{U}{}^{2}{}_{\sigma\mu} + R'_{V}{}^{1}{}_{\sigma\mu} + R'_{V}{}^{2}{}_{\sigma\mu} + R'_{Rflat}{}^{1}{}_{\sigma\mu} + R'_{Rflat}{}^{2}{}_{\sigma\mu} + R'_{\Omega}{}^{1}{}_{\sigma\mu} + R'_{\Omega}{}^{2}{}_{\sigma\mu} + R'_{B}{}^{1\beta}{}_{\sigma\beta\mu} +$$
$$+ R'_{B}{}^{2\beta}{}_{\sigma\beta\mu} + R^{1}{}_{\sigma\mu} + R^{2}{}_{\sigma\mu}$$
$$= R'^{1}{}_{\sigma\mu} + R'^{2}{}_{\sigma\mu} \tag{5.11}$$

where

$$R'^{1}{}_{\sigma\mu} = R'_{E}{}^{1}{}_{\sigma\mu} + R'_{SU(2)}{}^{1}{}_{\sigma\mu} + R'_{DE}{}^{1}{}_{\sigma\mu} + R'_{DSU(2)}{}^{1}{}_{\sigma\mu} + R'_{SU(3)}{}^{1}{}_{\sigma\mu} + R'_{U}{}^{1}{}_{\sigma\mu} + R'_{V}{}^{1}{}_{\sigma\mu} + R'_{Rflat}{}^{1}{}_{\sigma\mu} +$$
$$+ R'_{\Omega}{}^{1}{}_{\sigma\mu} + R'_{B}{}^{1\beta}{}_{\sigma\beta\mu} + R^{1}{}_{\sigma\mu} \tag{5.12}$$

$$R'^{2}{}_{\sigma\mu} = R'_{E}{}^{2}{}_{\sigma\mu} + R'_{SU(2)}{}^{2}{}_{\sigma\mu} + R'_{DE}{}^{2}{}_{\sigma\mu} + R'_{DSU(2)}{}^{2}{}_{\sigma\mu} + R'_{SU(3)}{}^{2}{}_{\sigma\mu} + R'_{U}{}^{2}{}_{\sigma\mu} + R'_{V}{}^{2}{}_{\sigma\mu} + R'_{Rflat}{}^{2}{}_{\sigma\mu} +$$
$$+ R'_{\Omega}{}^{2}{}_{\sigma\mu} + R'_{B}{}^{2\beta}{}_{\sigma\beta\mu} + R^{2}{}_{\sigma\mu}$$

with

$$R'_{E}{}^{1}{}_{\sigma\mu} = iF_{E}{}^{1}{}_{\sigma\mu}$$
$$R'_{E}{}^{2}{}_{\sigma\mu} = iF_{E}{}^{2}{}_{\sigma\mu}$$

$$R'_{SU(2)}{}^{1}{}_{\sigma\mu} = iF_{W}{}^{1}{}_{\sigma\mu}$$
$$R'_{SU(2)}{}^{2}{}_{\sigma\mu} = iF_{W}{}^{2}{}_{\sigma\mu}$$

$$R'_{DE}{}^{1}{}_{\sigma\mu} = iF_{DE}{}^{1}{}_{\sigma\mu}$$
$$R'_{DE}{}^{2}{}_{\sigma\mu} = iF_{DE}{}^{2}{}_{\sigma\mu}$$

$$R'_{DSU(2)}{}^{1}{}_{\sigma\mu} = iF_{DW}{}^{1}{}_{\sigma\mu}$$
$$R'_{DSU(2)}{}^{2}{}_{\sigma\mu} = iF_{DW}{}^{2}{}_{\sigma\mu}$$

$$R'_{SU(3)}{}^{1}{}_{\sigma\mu} = iF_{SU(3)}{}^{1}{}_{\sigma\mu}$$

$$R'_{SU(3)}{}^{2}{}_{\sigma\mu} = iF_{SU(3)}{}^{2}{}_{\sigma\mu}$$

$$R'_{U}{}^{1}{}_{\sigma\mu} = iF_{U}{}^{1}{}_{\sigma\mu}$$
$$R'_{U}{}^{2}{}_{\sigma\mu} = iF_{U}{}^{2}{}_{\sigma\mu}$$

$$R'_{V}{}^{1}{}_{\sigma\mu} = iF_{V}{}^{1}{}_{\sigma\mu}$$
$$R'_{V}{}^{2}{}_{\sigma\mu} = iF_{V}{}^{2}{}_{\sigma\mu}$$

$$R'_{Rflat}{}^{1}{}_{\sigma\mu} = iF_{Rflat}{}^{1}{}_{\sigma\mu}$$
$$R'_{Rflat}{}^{2}{}_{\sigma\mu} = iF_{Rflat}{}^{2}{}_{\sigma\mu}$$

$$R'_{\Omega}{}^{1}{}_{\sigma\mu} = iF_{\Omega}{}^{1}{}_{\sigma\mu}$$
$$R'_{\Omega}{}^{2}{}_{\sigma\mu} = iF_{\Omega}{}^{2}{}_{\sigma\mu}$$

$$R'_{B}{}^{1}{}_{\sigma\mu} = iB^{1}{}_{\sigma\mu}$$
$$R'_{B}{}^{2}{}_{\sigma\mu} = iB^{2}{}_{\sigma\mu}$$

with the further definition of $R''^{1}{}_{\sigma\mu}$ and $R''^{2}{}_{\sigma\mu}$:

$$R''^{1}{}_{\sigma\mu} = R'_{SU(3)}{}^{1}{}_{\sigma\mu} + R^{1}{}_{\sigma\mu}$$
$$R''^{2}{}_{\sigma\mu} = R'_{SU(3)}{}^{2}{}_{\sigma\mu} + R^{2}{}_{\sigma\mu} \tag{5.13}$$

Eq. 5.12 is the Ricci tensor. An additional Ricci-like tensor is

$$H_{\sigma\mu} = H^{\beta}{}_{\sigma\beta\mu} \tag{5.14}$$

The curvature scalar is

$$R' = g^{\sigma\mu}R'_{\sigma\mu} = + \partial^{\sigma}H^{1\beta}{}_{\sigma\beta} - \partial_{\beta}H^{1\beta}{}_{\sigma}{}^{\sigma} + H^{1\gamma}{}_{\beta\sigma}H^{1\beta}{}_{\gamma}{}^{\sigma} - H^{1\gamma}{}_{\mu\sigma}H^{1\beta}{}_{\gamma\beta} + \partial^{\sigma}H^{2\beta}{}_{\sigma\beta} - \partial_{\beta}H^{2\beta}{}_{\sigma}{}^{\sigma} +$$
$$+ H^{2\gamma}{}_{\beta\sigma}H^{2\beta}{}_{\gamma}{}^{\sigma} - H^{2\gamma\sigma}{}_{\sigma}H^{2\beta}{}_{\gamma\beta} + H^{1\gamma}{}_{\beta\sigma}H^{2\beta}{}_{\gamma}{}^{\sigma} - H^{1\gamma\sigma}{}_{\sigma}H^{2\beta}{}_{\gamma\beta} + H^{2\gamma}{}_{\beta\sigma}H^{1\beta}{}_{\gamma}{}^{\sigma} - H^{2\gamma\sigma}{}_{\sigma}H^{1\beta}{}_{\gamma\beta}$$

$$= g^{\sigma\mu}(R^{1\beta}{}_{\sigma\beta\mu} + R^{2\beta}{}_{\sigma\beta\mu}) \tag{5.15}$$

Additional curvature scalars are

$$H = g^{\sigma\mu}H_{\sigma\mu} \tag{5.16}$$
$$R'^{2} = g^{\sigma\mu}R'^{2}{}_{\sigma\mu} \tag{5.17}$$

# 6. Total Boson Lagrangian

In this chapter[23] we define a total vector boson and gravitational lagrangian as a generalization of the usual Einstein lagrangian with additional higher derivative terms added.

The major aspect of our extension is the introduction of higher derivative terms in a such a manner that they can be handled by canonical lagrangian methods to obtain the dynamical equations of motion and the equal time commutation relations (for the 'free' field approximations.)

We separate the Ricci tensor into two parts in order to use pseudoQuantization field theory (with fields labeled '1' and '2') to implement canonical lagrangian methods, and to introduce the flat space metric $\eta^{\sigma\mu}$ by a Higgs Mechanism.[24] The constant flat space metric part $\eta^{\sigma\mu}$ of the weak field quantum metric is usually an assumed quantity. But its close relation to the quantum field $g^{\sigma\mu}$ suggests that it could be generated by the same Higgs Mechanism that generates particle mass constants.[25]

We assume the lagrangian density:[26]

$$\mathfrak{L} = \mathrm{Tr}\,\sqrt{g}[MD_\nu R''^1{}_{\sigma\mu}D^\nu R''^{2\sigma\mu} + aR'^1{}_{\sigma\mu}R'^{2\sigma\mu} + bR' + cg^{\sigma\mu}g^2{}_{\sigma\mu} + c'g^{2\sigma\mu}g^2{}_{\sigma\mu} - dA_{SU(3)}{}^2{}_\mu A_{SU(3)}{}^{2\mu}] \tag{6.1}$$

where M, a, b, c, c', and d are constants to be determined later, and $R''^i{}_{\sigma\mu}$ for i = 1, 2 is determined by eq. 5.13.[27]

---

[23] Much of the material in this chapter appeared in Blaha (2016h) and earlier books.

[24] In earlier books such as Blaha (2016f) we showed that the use of two fields for each particle type enables us to clearly separate the 'vacuum expectation value' from its associated second quantized 'Higgs' field. The application to the weak field approximation for gravitons is one example.

[25] See also Blaha (2016c).

[26] Since the lagrangian terms are matrices it is necessary to take the trace.

[27] One may ask why $R''^1{}_{\sigma\mu}$ and $R''^2{}_{\sigma\mu}$ appear in the first term of the lagrangian, and not other interaction terms. We believe the primary reason is given in I "The extended vierbein $l^{\mu ai}(x)$ can be viewed as located at a point in a 32-dimensional complex-valued space.

$$l^{\mu ai}(x) = (\partial\xi_X{}^{ai}(x)/\partial x_\mu)_{X=h(x)} \tag{4.9}$$

where $\xi_X{}^{ai}$ is a set of locally inertial coordinates located at a 32-dimensional point X, and x = h(x) is a 4-dimensional point in a tangent subspace of the 32-dimensional space:

$$X = h(x)$$

The relation between complex 4-dimensional coordinates x and the 32-dimensional coordinates X is an embedding of a 4-dimensional surface within a 32-dimensional complex space when account is taken of the range of possible x values. We have considered such embeddings in Blaha (2015a), and in earlier books, and developed a theory of a 16-dimensional complex-valued space (the *Megaverse*) that contains our universe and probably many other universes." Thus SU(3) and Gravitation have a special role in our particle dynamics based on geometry. The secondary, practical

This higher derivative lagrangian maintains the locality of the theory but does entail a modest modification in the derivation of the Euler-Lagrange equations of motion. It also requires the use of principal value propagators rather than ordinary Feynman propagators for gluon and graviton interactions. Thus the Strong Interaction sector, and the Gravitation sector are Action-at-a-Distance theories that are similar in spirit to Wheeler-Feynman Electrodynamics. The two U(1) Electromagnetic sectors, the Generation group U(4) gauge field sector, the Layer group U(4) gauge field sector, the two SU2) Weak sectors, the U(4) 'Rflat' gauge field sector, the spinor connection sector, and the Ω-interaction sector may, or may not, be Action-at-a-Distance fields. They are not constrained to be Action-at-a-Distance by the present considerations.

Since we wish to apply it cosmologically, and within hadrons, where the gravitational spinor connections are negligible due to the smallness of the gravitational constant G and the 'smallness' of B spin on the cosmological scale, we set ${B^1}_{\nu\mu} = {B^2}_{\nu\mu} = 0$ and find[28]

$$\begin{aligned}\mathcal{L} = \mathrm{Tr}\ \sqrt{g}[MD_\nu({R'^1}_{SU(3)\sigma\mu} + {R'_G}^1{}_{\sigma\mu})D^\nu({R'_{SU(3)}}^{2\sigma\mu} + {R'_G}^{2\sigma\mu}) + \\ + a{R'^1}_{\sigma\mu}R'^{2\sigma\mu} + bR' + cg^{\sigma\mu}{g^2}_{\sigma\mu} + c'g^{2\sigma\mu}{g^2}_{\sigma\mu} - d{A_{SU(3)}}^2{}_\mu {A_{SU(3)}}^{2\mu}]\end{aligned} \tag{6.2}$$

Since there are no strong interaction fields in 'empty' space and gravity is negligible within hadrons,[29] we can drop the interaction terms between these interactions. However, we cannot drop the interaction terms amongst Electromagnetism, the Weak interaction, the Strong Interaction, the Generation group U(4) interaction, the Layer group U(4) interaction, the 'Rflat' U(4) group interaction, and the SU(3)⊗U(64) Ω-interaction – within, and between, hadrons. The interaction terms between Electromagnetism and Gravitation are important cosmologically.

Eq. 6.2 can therefore be expressed as:[30]

$$\mathcal{L} = \mathcal{L}_E + \mathcal{L}_{SU(2)} + \mathcal{L}_{DE} + \mathcal{L}_{DSU(2)} + \mathcal{L}_{SU(3)} + \mathcal{L}_U + \mathcal{L}_V + \mathcal{L}_{Rflat} + \mathcal{L}_\Omega + \mathcal{L}_G + \mathcal{L}_{int} \tag{6.3}$$

where taking traces of $\mathcal{L}$s terms is understood

$$\begin{aligned}
\mathcal{L}_E &= \mathrm{Tr}\ \sqrt{g}\{M\{[\partial_\nu + i({A_E}^1{}_\nu + {A_E}^2{}_\nu)]{F^1}_{E\sigma\mu}[\partial^\nu + i({A_E}^{1\nu} + {A_E}^{2\nu})]{F^2}_E{}^{\sigma\mu}\} + a{F_E}^1{}_{\sigma\mu}{F_E}^{2\sigma\mu}\} \qquad (6.4)\\
\mathcal{L}_{SU(2)} &= \mathrm{Tr}\ \sqrt{g}[a{F_W}^1{}_{\sigma\mu}{F_W}^{2\sigma\mu}]\\
\mathcal{L}_{DE} &= \mathrm{Tr}\ \sqrt{g}\{M\{[\partial_\nu + i({A_{DE}}^1{}_\nu + {A_{DE}}^2{}_\nu)]{F^1}_{DE\sigma\mu}[\partial^\nu + i({A_{DE}}^{1\nu} + {A_{DE}}^{2\nu})]{F_{DE}}^{2\sigma\mu}\} + a{F_{DE}}^1{}_{\sigma\mu}{F_{DE}}^{2\sigma\mu}\}\\
\mathcal{L}_{DSU(2)} &= \mathrm{Tr}\ \sqrt{g}[a{F_W}^1{}_{\sigma\mu}{F_W}^{2\sigma\mu}]\\
\mathcal{L}_{SU(3)} &= \mathrm{Tr}\ \sqrt{g}\{M[\partial_\nu + i({A_{SU(3)}}^1{}_\nu + {A_{SU(3)}}^2{}_\nu)]{F_{SU(3)}}^1{}_{\sigma\mu}[\partial^\nu + i({A_{SU(3)}}^{1\nu} + {A_{SU(3)}}^{2\nu})]{F_{SU(3)}}^{2\sigma\mu} +
\end{aligned}$$

---

reason is the experimental reality that the Strong Interaction and Gravitation are known to have 'anomalous' features that will be seen to be remedied by these insertions while the other interactions are 'conventional.'

[28] The constants have the dimensions: M has the dimension of inverse mass squared, b has dimension mass squared, a is dimensionless, c and c' have dimension mass to the 4th order, and d has dimension mass squared.

[29] We show gravity weakens at very short distances using our Two-Tier Quantum Field Theory formalism. See Blaha (2003) and (2005a) amog other books by the author.

[30] We only consider the gauge field lagrangian terms in this chapter and in this book.

$$+ aF_{SU(3)}{}^{1}{}_{\sigma\mu}F_{SU(3)}{}^{2\sigma\mu} - dA_{SU(3)}{}^{2}{}_{\mu}A_{SU(3)}{}^{2\mu}\}$$

$$\mathcal{L}_U = \text{Tr}\,\surd g[aF_U{}^{1}{}_{\sigma\mu}F_U{}^{2\sigma\mu}]$$

$$\mathcal{L}_V = \text{Tr}\,\surd g[aF_V{}^{1}{}_{\sigma\mu}F_V{}^{2\sigma\mu}]$$

$$\mathcal{L}_{Rflat} = \text{Tr}\,\surd g[aF_{Rflat}{}^{1}{}_{\sigma\mu}F_{Rflat}{}^{2\sigma\mu}]$$

$$\mathcal{L}_\Omega = \text{Tr}\,\surd g[aF_\Omega{}^{1}{}_{\sigma\mu}F_\Omega{}^{2\sigma\mu}]$$

$$\mathcal{L}_G = \text{Tr}\,\surd g[MD_\nu R^{1}{}_{\sigma\mu}D^\nu R^{2\sigma\mu} + aR^{1}{}_{\sigma\mu}R^{2\sigma\mu} + bg^{\sigma\mu}(R^{1\beta}{}_{\sigma\beta\mu} + R^{2\beta}{}_{\sigma\beta\mu}) + cg^{\sigma\mu}g^{2}{}_{\sigma\mu} + c'g^{2\sigma\mu}g^{2}{}_{\sigma\mu}]$$

$$= \text{Tr}\,\surd g[MD_\nu R^{1}{}_{\sigma\mu}D^\nu R^{2\sigma\mu} + aR^{1}{}_{\sigma\mu}R^{2\sigma\mu} + bH + cg^{\sigma\mu}g^{2}{}_{\sigma\mu} + c'g^{2\sigma\mu}g^{2}{}_{\sigma\mu}]$$

$$\mathcal{L}_{int} = \mathcal{L} - (\mathcal{L}_E + \mathcal{L}_{SU(2)} + \mathcal{L}_{DE} + \mathcal{L}_{DSU(2)} + \mathcal{L}_{SU(3)} + \mathcal{L}_U + \mathcal{L}_V + \mathcal{L}_{Rflat} + \mathcal{L}_\Omega + \mathcal{L}_G)$$

Thus $\mathcal{L}_{SU(3)}$, $\mathcal{L}_{SU(2)}$, $\mathcal{L}_E$ , $\mathcal{L}_{DE}$, $\mathcal{L}_{DSU(2)}$, $\mathcal{L}_U$,, $\mathcal{L}_V$, $\mathcal{L}_{Rflat}$, $\mathcal{L}_\Omega$, and parts of $\mathcal{L}_{int}$ are the dominant interactions within hadrons, and $\mathcal{L}_G$, $\mathcal{L}_E$ and parts of $\mathcal{L}_{int}$ are the dominant interactions in space within the framework of this discussion.

The $D_\nu R^{1}{}_{\sigma\mu}$ and $D^\nu R^{2\sigma\mu}$ terms have the form:

$$D_\nu R^{i}{}_{\sigma\mu} = + \partial_\nu R^{i}{}_{\sigma\mu} \;\; - H^{1\beta}{}_{\sigma\nu}R^{i}{}_{\beta\mu} - H^{1\beta}{}_{\nu\mu}R^{i}{}_{\sigma\beta}$$

for i = 1, 2.

## 6.1 The Eleven Interaction Unified Theory

*The unification implemented in an eleven interaction Riemann-Christoffel curvature tensor leads to new interaction terms beyond those in The Standard Model. These additional interactions in $\mathcal{L}_{int}$ imply new phenomena in this unified theory such as: 1) a possible relationship between the various coupling constants; 2) a possible relationship between parts of these interactions; 3) a possible explanation of the deviations of gravity at intra-galactic distances and inter-galactic distances; 4) a possible solution of the proton spin puzzle due to electromagnetic-gluon terms in $\mathcal{L}_{int}$ that have not been previously considered; and 5) a possible explanation of the proton radius puzzle found in experiment due to Generation interaction-Electromagnetic interaction terms in $\mathcal{L}_{int}$, as well as other results.*

## 6.2 The Strong Interaction Sector

The Strong Interaction lagrangian density terms are:[31,32]

$$\mathcal{L}_{SU(3)} = \text{Tr}\,\surd g[MD_{SU(3)\nu}R'^{1}{}_{SU(3)\sigma\mu}D_{SU(3)}{}^{\nu}R'^{2}{}_{SU(3)}{}^{\sigma\mu} + aR'^{1}{}_{SU(3)\sigma\mu}R'^{2}{}_{SU(3)}{}^{\sigma\mu} - dA_{SU(3)}{}^{2}{}_{\mu}A_{SU(3)}{}^{2\mu}] \tag{6.5}$$

with

[31] The form is virtually identical to S. Blaha, Phys. Rev. D**11**, 2921 (1974), and Blaha's 1976 Gravity Research Foundation Essay, except for the initial derivative term. See Appendices A and D.

[32] We note the constant a, which appears in this chapter and chapter 5 is NOT the Charmonium constant a.

$$D_{SU(3)\nu} = [\partial_\nu + if(A_{SU(3)}{}^1{}_\nu + A_{SU(3)}{}^2{}_\nu)]$$

and the electromagnetic lagrangian density term is now

$$\mathcal{L}_E = \sqrt{g}\{aF_E{}^1{}_{\sigma\mu}F_E{}^{2\sigma\mu}\} \tag{6.6}$$

Corresponding changes take place in $\mathcal{L}_{SU(2)}$ and $\mathcal{L}_{int}$.

We now approximate the metric determinant as g = 1 within hadrons. Thus the Strong lagrangian part becomes

$$\mathcal{L}_{SU(3)} = \text{Tr}\ \{MD_\nu R'^1{}_{SU(3)\sigma\mu}D^\nu R'^2{}_{SU(3)}{}^{\sigma\mu} + aR'^1{}_{SU(3)\sigma\mu}R'^2{}_{SU(3)}{}^{\sigma\mu} - dA_{SU(3)}{}^2{}_\mu A_{SU(3)}{}^{2\mu}\} \tag{6.7}$$

$$= \text{Tr}\ \{MD_\nu R'^1{}_{SU(3)\sigma\mu}D^\nu R'^2{}_{SU(3)}{}^{\sigma\mu} + \zeta R'^1{}_{SU(3)\sigma\mu}R'^2{}_{SU(3)}{}^{\sigma\mu} - \varsigma A_{SU(3)}{}^2{}_\mu A_{SU(3)}{}^{2\mu}\} \tag{6.8}$$

where

$$D_\nu R'^j{}_{SU(3)\sigma\mu} = \partial_\nu R'^j{}_{SU(3)\sigma\mu} + g_{SU(3)}[A_{SU(3)}{}^1{}_\nu, R'^j{}_{SU(3)\sigma\mu}] \tag{6.9}$$

for j = 1, 2.

# 7. Gravitational Potential on the Three Distance Scales

This chapter and the next chapter describe some of the possible results of the unified eleven interaction theory described earlier. The discussions in this book and I focus almost entirely on the boson sector of a Theory of Everything. The fermion sector is described in detail in Blaha (2015a).

In this chapter we pull together results on the gravitational potential found in I. We note that a new experimental study of 33,000 galaxies indicated that the gravitational potential at inter-galactic distances deviates significantly from the Newtonian potential G/r. In I we showed that such a deviation occurs in our theory. This experimental result was not known at the time of the writing of I.[33]

## 7.1 Total Gravitational Potential

The gravitational potential generated from the real-valued affine connection term bR' in the lagrangian eq. 6.1 was shown in section 6.3 of I (with a massless graviton term required) to be

$$V^{tot}{}_G(\mathbf{r}) = -G/r - a_1Ge^{-m_Gr}/r + a_2G\cos(m_Gr)/r \qquad (6.25)$$

where

$$m_G{}^2 = 2b/a \cong 10^{55}\ ev^2 = 10^{27}\ GeV^2 \qquad (6.25a)$$
$$a_1 = ½$$
$$a_2 = ½$$

The gravitational potential contribution due to the gravitational gauge field $A_{Rflat}{}^{1\mu}$ is (according to I):

$$V_{Rflat}(\mathbf{r}) \equiv V_{GA1}(\mathbf{r}) = -[1/(96\pi a)][1/r - e^{-m_Ar}/r] \qquad (13.15)$$

where

$$m_A = (a/M)^{½} = m_{SI} \cong 10^{-71}\ GeV \qquad (13.16)$$

by eq. 5.20 of I. Since a ≅ 1 the coupling constant

$$1/(96\pi a) \cong 0.0033 \qquad (13.17)$$

In comparison the electromagnetic fine structure constant is

---

[33] The gravitational potential was found to be greater than G/r at inter-galactic distances in a survey of 33,000 galaxies by M. Brouwer and colleagues at the Leiden Observatory (The Netherlands) in an announcement on December 18, 2016.

$$\alpha \cong 0.0073$$

Thus the $A_R$ coupling constant is approximately ½ of the fine structure constant.

The total gravitational potential due to both sources of gravity is

$$V(\mathbf{r}) = -G/r - a_1 G e^{-m_G r}/r + a_1 G\cos(m_G r)/r - [1/(96\pi a)][1/r - e^{-m_A r}/r] \tag{7.1}$$

We now examine $V(\mathbf{r})$ at short distances (within the solar system), distances of tens of thousands of light years (intra-galactic distances), and distances between galaxies (hundreds of thousands to millions of light years and beyond).

## 7.2 Intra-Solar System Distance Scale

Since $m_G$ is very large and $m_A$ is extremely small, the gravitational potential at distances of up to at least several light years is approximately

$$V(\mathbf{r}) \cong -G/r \tag{7.2}$$

by eq. 7.1 to well within feasible experimental limits.

## 7.3 Galactic Distance Scale

At distances of several tens of thousands of light years up to perhaps 100,000 light years we find eq. 7.1 becomes approximately

$$V(\mathbf{r}) \cong -[G + 1/(96\pi a)]/r + \tfrac{1}{2}G\cos(m_G r)/r \tag{7.3}$$

Note that the approximate expansion of the terms in eq. 7.3 to third order yields

$$V(\mathbf{r}) \sim -[G + 1/(96\pi a)]/r + a_1 G m_G^3 r^2 - a_1 G m_G^2 r + \text{constants} \tag{7.4}$$

with the resultant force

$$\mathbf{F} = \nabla V^{tot}{}_G(\mathbf{r}) \sim [G + 1/(96\pi a)]\mathbf{r}/r^3 + 2a_1 G m_G^3 \mathbf{r} - a_2 G m_G^2 \mathbf{r}/r + \ldots \tag{7.5}$$

This result is to be compared to the MoND force of A. Balakin et al, Phys. Rev. **D70**, 064027 (2004):[34]

$$F = -\lambda G m[M/r^2 \; - |\,\Pi_c\,|\, r/c^2] \tag{7.6}$$

and the vector form suggested by H-S Zhao et al, Phys. Rev. **D82**, 103001 (2010):

[34] The constant c in eq. 7.6 is the speed of light, and M is the mass (not the M used in our lagrangian equations.)

$$\partial\Phi/\partial\mathbf{r} \;=\; \mathrm{Gm}\mathbf{r}/\mathrm{r}^3 + (\mathrm{Gm})^{1/2}\mathbf{r}/\mathrm{r}^2 \tag{7.7}$$

Recent studies of 153 galaxies confirm the MoND discrepancy from Newtonian gravitation.[35]

## 7.4 Intergalactic Distance Scale

We can estimate the gravitational potential of V(**r**) in eq. 7.1 for large distances of the order of many hundreds of thousands of light years, and beyond, we find

$$V(\mathbf{r}) \cong -[G + 1/(96\pi a)]/r \tag{7.8}$$

to good approximation since $m_G^{\,2} = 2b/a \cong 10^{55}$ ev$^2$ = $10^{27}$ GeV$^2$ sets a distance scale of the order of tens of thousands of light years causing the oscillating term to 'wash out,' and the $a_1Ge^{-m_Gr}/r$ term to be negligible. The $e^{-m_Ar}/r$ term whose distance scale is short range is also negligible.

*Consequently we find a deeper potential and thus a larger attractive gravitational force at inter-galactic distances in agreement with the 33,000 galaxy survey of M. Brouwer and colleagues.*

## 7.5 Theoretical Agreement With Gravitational Data at all Known Distances

*Our unified theory of the eleven interactions agrees with known gravitational data.*

[35] S. S. McGaugh, F. Lelli, and J. M. Schombert, arXiv: 1609.0591 (2016).

# 8. Other Effects of New Interactions Between Boson Interactions

## 8.1 Quark Confinement and the Charmonium Potential

Charmonium bound state studies[36] have yielded a gluon potential with the form

$$V(r) = -\kappa/r + r/a^2 \tag{8.1}$$

where $\kappa = 0.61$, $a = 2.38\ \text{GeV}^{-1}$ and the charmed quark mass was 1.84 GeV. In I we showed that this potential follows directly from our lagrangian in I and the more general lagrangian presented in chapter 6.

We conclude we have a viable Strong interaction subsector of our eleven interaction lagrangian. The charmonium potential emerges directly from our theory. We note that $g_{SU(3)}{}^2/4\pi = 0.024$ is only a factor of 3.3 more than the fine structure constant – approximately 1/137 = 0.0073. *Therefore perturbative corrections to our inter-quark potential may not be significant and our theory may be the correct theory of the strong interaction. The strong interaction potential in the charmonium fit suggests that the unperturbed potential of the theory may be a good approximation to the exact potential determined in perturbation theory.*

*The smallness of the Strong Interaction terms in the Cornell group potential looks puzzling at first glance. Why is it not large ("Strong")? We believe the strength of the Strong Interaction does not originate in the value of the coupling constant but rather in the linear potential term which provides confinement. The Cornell group potential's 'small' coupling constant then becomes understandable within the context of Strong Interaction phenomenology.*

## 8.2 Missing Nucleon Spin Puzzle

The estimates of nucleon spin that are obtained from parton analyses of deep inelastic electron – nucleon interactions are woefully short of the spin expected in quark models of nucleons. The missing spin has been attributed to a number of causes. However the Missing Spin Puzzle remains.

From eq. 6.3-6.4 it is clear that there are important new interaction terms between the Electromagnetic and Strong interaction fields. After taking traces we find

$$\mathcal{L}_{intEM\text{-}S} = -\text{Tr}\ iM\{(A_E{}^1{}_\nu + A_E{}^2{}_\nu)\, F_{SU(3)}{}^1{}_{\sigma\mu} D^\nu F_{SU(3)}{}^{2\sigma\mu} + iD_\nu F_{SU(3)}{}^1{}_{\sigma\mu}(A_E{}^{1\nu} + A_E{}^{2\nu}) F_{SU(3)}{}^{2\sigma\mu}\} \tag{8.2}$$

[36] This chapter first appeared in Blaha (2016a).

$\mathcal{L}_{\text{intEM-S}}$ generates a combined photon-gluon vertex insertion in gluon interactions between quarks within a hadron. Figs. 8.1 and 8.2 show two simple possible vertex insertions in a gluon line.

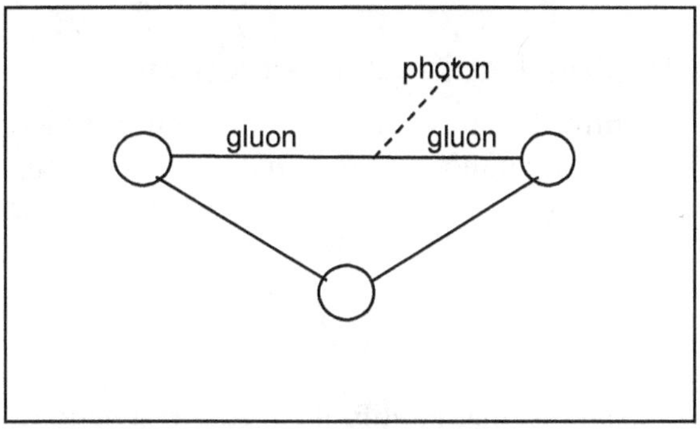


Figure 8.1. A single 'outgoing' photon vertex insertion in a gluon line. Only single gluon lines between the three quarks are displayed.

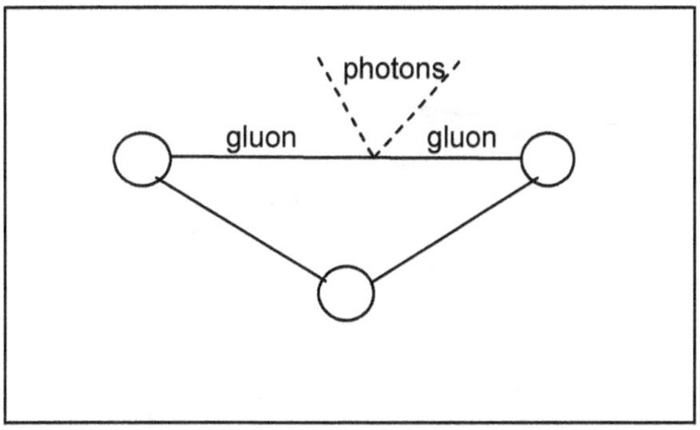


Figure 8.2. An 'outgoing' two photon vertex insertion in a gluon line. Only single gluon lines between the three quarks are displayed.

The gluon line, by itself, has $1/k^4$ and $1/k^2$ momentum space propagator terms. *The insertion of the vertex in Fig.8.1 generated by* $\mathcal{L}_{intEM\text{-}S}$ *yields a combined momentum factor of* $k^3(k^4k^4)^{-1} = k^{-5}$ *which would make it (summed over all gluon lines) a significant contribution to the proton spin determination in deep inelastic electron-nucleon scattering.*[37] *The insertion of the vertex in Fig. 8.2 generated by* $\mathcal{L}_{intEM\text{-}S}$ *yields a combined momentum factor of* $k^2(k^4k^4)^{-1} = k^{-6}$ *which may have a less significant effect.*

---

[37] See C. A. Aidala, S. D. Bass, D. Hasch, and G. K. Mallot, arXiv: 1209.2803v2 (2013) and references therein for a review of the 'missing' nucleon spin puzzle.

*Thus our unified theory may solve the nucleon spin puzzle. The interactions in Figs. 8.1 and 8.2 introduce a new direct connection between photons and spin one gluons. Thus their contributions to the summations of proton spin interactions in parton models may account for the 'missing' two-thirds of proton spin. Our unified theory has a new gluon-photon interaction that is not found in the conventional Standard Model.*

## 8.3 Discrepancies between Proton Radius Measurements

Recently experiments have confirmed that the radius of a proton in a muonic hydrogen atom is smaller than the proton radius measured in a conventional hydrogen atom composed of a proton encircled by an electron. The lagrangian in eq. 6.1 indicates that there are direct interactions between photons and Generation group gauge fields such as

$$A_E{}^1 A_E{}^{2\nu} U^{1i}{}_\nu U^{2i\nu}$$

These interactions result in Feynman diagrams that modify the electromagnetic field between a proton and a circling muon or electron. Fig. 8.3 shows the simplest forms of this interaction for a muon and a quark within a proton.

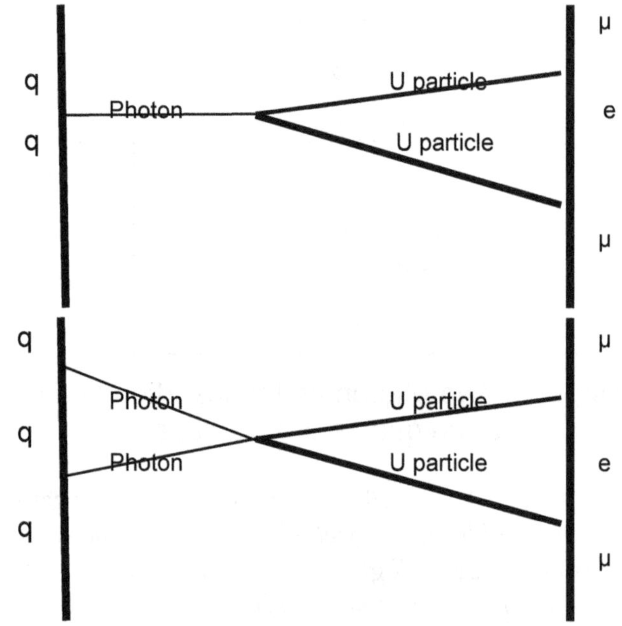


Figure 8.3. Photon-U gauge particle interactions between a muon and quark (within a proton).

The net effect of this additional interaction is to increase the force between the quark and muon beyond what one would expect using the electromagnetic potential, from the usual

$$V = -\,e^2/r$$

to

$$V' = -\,(e + \delta)^2/r \qquad (8.3)$$
$$= -\,e^2/r'$$

where δ is small compared to e, and where

$$r' = r[e/(e + \delta)]^2 < r \qquad (8.4)$$

giving an apparently smaller radial distance than the actual muon radial distance. This change would cause an energy shift to a lower value (more negative) in the muonic energy levels and, consequently, the proton radius would appear smaller for muonic atoms – as experiment confirms.[38]

Thus the new 'interactions between interactions' of our unified theory have significant effects that are already qualitatively verified by experiment.

[38] Fig. 8.3 is not of importance for 'normal' hydrogen since the electron is so much lighter than the muon.

# REFERENCES


Bjorken, J. D., Drell, S. D., 1964, *Relativistic Quantum Mechanics* (McGraw-Hill, New York, 1965).

Bjorken, J. D., Drell, S. D., 1965, *Relativistic Quantum Fields* (McGraw-Hill, New York, 1965).

Blaha, S., 1998, *Cosmos and Consciousness* (Pingree-Hill Publishing, Auburn, NH, 1998).

_______, 2002, *A Finite Unified Quantum Field Theory of the Elementary Particle Standard Model and Quantum Gravity Based on New Quantum Dimensions™ & a New Paradigm in the Calculus of Variations* (Pingree-Hill Publishing, Auburn, NH, 2002).

_______, 2003, *A Finite Unified Quantum Field Theory of the Elementary Particle Standard Model and Quantum Gravity Based on New Quantum Dimensions™ and a New Paradigm in the Calculus of Variations* (Pingree-Hill Publishing, Auburn, NH, 2003).

_______, 2004, *Quantum Big Bang Cosmology: Complex Space-time General Relativity, Quantum Coordinates™ Dodecahedral Universe, Inflation, and New Spin 0, ½, 1 & 2 Tachyons & Imagyons* (Pingree-Hill Publishing, Auburn, NH, 2004).

______, *2005a, Quantum Theory of the Third Kind: A New Type of Divergence-free Quantum Field Theory Supporting a Unified Standard Model of Elementary Particles and Quantum Gravity based on a New Method in the Calculus of Variations* (Pingree-Hill Publishing, Auburn, NH, 2005).

______, 2005b, *The Metatheory of Physics Theories, and the Theory of Everything as a Quantum Computer Language* (Pingree-Hill Publishing, Auburn, NH, 2005).

______, 2005c, *The Equivalence of Elementary Particle Theories and Computer Languages: Quantum Computers, Turing Machines, Standard Model, Superstring Theory, and a Proof that Gödel's Theorem Implies Nature Must Be Quantum* (Pingree-Hill Publishing, Auburn, NH, 2005).

______, 2006a, *The Foundation of the Forces of Nature* (Pingree-Hill Publishing, Auburn, NH, 2006).

______, 2006b, *A Derivation of ElectroWeak Theory based on an Extension of Special Relativity; Black Hole Tachyons; & Tachyons of Any Spin.* (Pingree-Hill Publishing, Auburn, NH, 2006).

_______, 2007a, *Physics Beyond the Light Barrier: The Source of Parity Violation, Tachyons, and A Derivation of Standard Model Features* (Pingree-Hill Publishing, Auburn, NH, 2007).

______, 2007b, *The Origin of the Standard Model: The Genesis of Four Quark and Lepton Species, Parity Violation, the ElectroWeak Sector, Color SU(3), Three Visible Generations of Fermions, and One Generation of Dark Matter with Dark Energy* (Pingree-Hill Publishing, Auburn, NH, 2007).

______, *2008a, A Direct Derivation of the Form of the Standard Model From GL(16) (Pingree-Hill Publishing, Auburn, NH, 2008).*

______, 2008b, *A Complete Derivation of the Form of the Standard Model With a New Method to Generate Particle Masses Second Edition* (Pingree-Hill Publishing, Auburn, NH, 2008)

______, 2009, *The Algebra of Thought & Reality: The Mathematical Basis for Plato's Theory of Ideas, and Reality Extended to Include A Priori Observers and Space-Time Second Edition* (Pingree-Hill Publishing, Auburn, NH, 2009).

______, 2010a, *Operator Metaphysics: A New Metaphysics Based on a New Operator Logic and a New Quantum Operator Logic that Lead to a Mathematical Basis for Plato's Theory of Ideas and Reality* (Pingree-Hill Publishing, Auburn, NH, 2010).

______, 2010b, *The Standard Model's Form Derived from Operator Logic, Superluminal Transformations and GL(16)* (Pingree-Hill Publishing, Auburn, NH, 2010).

______, 2011a, *21st Century Natural Philosophy Of Ultimate Physical Reality* (McMann-Fisher Publishing, Auburn, NH, 2011).

______, 2011b, *All the Universe! Faster Than Light Tachyon Quark Starships & Particle Accelerators with the LHC as a Prototype Starship Drive Scientific Edition* (Pingree-Hill Publishing, Auburn, NH, 2011).

______, 2011c, *From Asynchronous Logic to The Standard Model to Superflight to the Stars* (Blaha Research, Auburn, NH, 2011).

______, 2012a, *From Asynchronous Logic to The Standard Model to Superflight to the Stars volume 2: Superluminal CP and CPT, U(4) Complex General Relativity and The Standard*

*Model, Complex Vierbein General Relativity, Kinetic Theory, Thermodynamics* (Blaha Research, Auburn, NH, 2012).

______, 2012b, *Standard Model Symmetries, And Four And Sixteen Dimension Complex Relativity; The Origin Of Higgs Mass Terms* (Blaha Reasearch, Auburn, NH, 2012).

______, 2013a, *Multi-Stage Space Guns, Micro-Pulse Nuclear Rockets, and Faster-Than-Light Quark-Gluon Ion Drive Starships* (Blaha Research, Auburn, NH, 2013).

______, 2013b, *The Bridge to Dark Matter; A New Sister Universe; Dark Energy; Inflatons; Quantum Big Bang; Superluminal Physics; An Extended Standard Model Based on Geometry* (Blaha Reasearch, Auburn, NH, 2013).

______, 2014a, *Universes and Multiverses: From a New Standard Model to a Physical Multiverse; The Big Bang; Our Sister Universe's Wormhole; Origin of the Cosmological Constant, Spatial Asymmetry of the Universe, and its Web of Galaxies; A Baryonic Field between Universes and Particles; Flatverse Extended Wheeler-DeWitt Equation* (Blaha Reasearch, Auburn, NH, 2014).

______, 2014b, *All the Multiverse! Starships Exploring the Endless Universes of the Cosmos Using the Baryonic Force* (Blaha Research, Auburn, NH, 2014).

______, 2014c, *All the Multiverse! II Between Multiverse Universes: Quantum Entanglement Explained by the Multiverse Coherent Baryonic Radiation Devices – PHASERs Neutron Star Multiverse Slingshot Dynamics Spiritual and UFO Events, and the Multiverse Microscopic Entry into the Multiverse* (Blaha Research, Auburn, NH, 2014).

______, 2015a, *PHYSICS IS LOGIC PAINTED ON THE VOID: Origin of Bare Masses and The Standard Model in Logic, U(4) Origin of the Generations, Normal and Dark Baryonic Forces, Dark Matter, Dark Energy, The Big Bang, Complex General Relativity, A Megaverse of Universe Particles* (Blaha Research, Auburn, NH, 2015).

______, 2015b, *PHYSICS IS LOGIC Part II: The Theory of Everything, The Megaverse Theory of Everything, U(4)⊗U(4) Grand Unified Theory (GUT), Inertial Mass = Gravitational Mass, Unified Extended Standard Model and a New Complex General Relativity with Higgs Particles, Generation Group Higgs Particles* (Blaha Research, Auburn, NH, 2015).

______, 2015c, *The Origin of Higgs ("God") Particles and the Higgs Mechanism: Physics is Logic III, Beyond Higgs – A Revamped Theory With a Local Arrow of Time, The Theory of Everything Enhanced, Why Inertial Frames are Special, Universes of the Mind* (Blaha Research, Auburn, NH, 2015).

______, 2015d, *The Origin of the Eight Coupling Constants of The Theory of Everything: U(8) Grand Unified Theory of Everything (GUTE), $S^8$ Coupling Constant Symmetry, Space-Time Dependent Coupling Constants, Big Bang Vacuum Coupling Constants, Physics is Logic IV* (Blaha Research, Auburn, NH, 2015).

______, 2016a, *New Types of Dark Matter, Big Bang Equipartition, and A New U(4) Symmetry in the Theory of Everything: Equipartition Principle for Fermions, Matter is 83.33% Dark, Penetrating the Veil of the Big Bang, Explicit QFT Quark Confinement and Charmonium, Physics is Logic V* (Blaha Research, Auburn, NH, 2016).

______, 2016b, *The Periodic Table of the 192 Quarks and Leptons in The Theory of Everything: The U(4) Layer Group, Physics is Logic VI* (Blaha Research, Auburn, NH, 2016).

______, 2016c, *New Boson Quantum Field Theory, Dark Matter Dynamics, Dark Matter Fermion Layer Mixing, Genesis of Higgs Particles, New Layer Higgs Masses, Higgs Coupling Constants, Non-Abelian Higgs Gauge Fields, Physics is Logic VII* (Blaha Research, Auburn, NH, 2016).

______, 2016d, *Unification of the Strong Interactions and Gravitation: Quark Confinement Linked to Modified Short-Distance Gravity; Physics is Logic VIII* (Blaha Research, Auburn, NH, 2016).

______, 2016e, *MoND: Unification of the Strong Interactions and Gravitation II, Quark Confinement Linked to Large-Scale Gravity, Physics is Logic IX* (Blaha Research, Auburn, NH, 2016).

______, 2016f, *CQMechanics: A Unification of Quantum & Classical Mechanics, Quantum/Semi-Classical Entanglement, Quantum/Classical Path Integrals, Quantum/Classical Chaos* (Blaha Research, Auburn, NH, 2016).

______, 2016g, *GEMS: Unified Gravity, ElectroMagnetic and Strong Interactions: Manifest Quark Confinement, A Solution for the Proton Spin Puzzle, Modified Gravity on the Galactic Scale* (Pingree Hill Publishing, Auburn, NH, 2016).

______, 2016h, *Unification of the Seven Boson Interactions based on the Riemann-Christoffel Curvature Tensor* (Pingree Hill Publishing, Auburn, NH, 2016).

Chrystal, G., 1961, *Textbook of Algebra Part One* (Dover Publications, Inc., New York, 1961).

Eddington, A. S., 1952, *The Mathematical Theory of Relativity* (Cambridge University Press, Cambridge, U.K., 1952).

Fant, Karl M., 2005, *Logically Determined Design: Clockless System Design With NULL Convention Logic* (John Wiley and Sons, Hoboken, NJ, 2005).
Heitler, W., 1954, *The Quantum Theory of Radiation* (Claendon Press, Oxford, UK, 1954).

Huang, Kerson, 1992, *Quarks, Leptons & Gauge Fields 2nd Edition* (World Scientific Publishing Company, Singapore, 1992).

Misner, C. W., Thorne, K. S., and Wheeler, J. A., 1973, *Gravitation* (W. H. Freeman, New York, 1973).

Sagan, H., 1993, *Introduction to the Calculus of Variations* (Dover Publications, Mineola, NY, 1993).

Sakurai, J. J., 1964, *Invariance Principles and Elementary Particles* (Princeton University Press, Princeton, NJ, 1964).

Streater, R. F. and Wightman, A. S., 2000, *PCT, Spin, Statistics, and All That* (Princeton University Press, Princeton, NJ 2000).

Weinberg, S., 1972, *Gravitation and Cosmology* (John Wiley and Sons, New York, 1972).

Weinberg, S., 1995, *The Quantum Theory of Fields Volume I* (Cambridge University Press, New York, 1995).

Weyl, H., 1950, *Space, Time, Matter* (Dover, New York, 1950).

Weyl, H., (Tr. S. Pollard et al), 1987, *The Continuum* (Dover Publications, New York, 1987).

# INDEX

# About the Author

Stephen Blaha is a well known Physicist and Man of Letters with interests in Science, Society and civilization, the Arts, and Technology. He had an Alfred P. Sloan Foundation scholarship in college. He received his Ph.D. in Physics from Rockefeller University. He has served on the faculties of several major universities. He was also a Member of the Technical Staff at Bell Laboratories, a manager at the Boston Globe Newspaper, a Director at Wang Laboratories, and President of Blaha Software Inc and of Janus Associates Inc. (NH).

Among other achievements he was a co-discoverer of the "r potential" for heavy quark binding developing the first (and still the only demonstrable) non-abelian gauge theory with an "r" potential; first suggested the existence of topological structures in superfluid He-3; first proposed Yang-Mills theories would appear in condensed matter phenomena with non-scalar order parameters; first developed a grammar-based formalism for quantum computers and applied it to elementary particle theories; first developed a new form of quantum field theory without divergences (thus solving a major 60 year old problem that enabled a unified theory of the Standard Model and Quantum Gravity without divergences to be developed); first developed a formulation of complex General Relativity based on analytic continuation from real space-time; first developed a generalized non-homogeneous Robertson-Walker metric that enabled a quantum theory of the Big Bang to be developed without singularities at $t = 0$; first generalized Cauchy's theorem and Gauss' theorem to complex, curved multi-dimensional spaces; received Honorable Mention in the Gravity Research Foundation Essay Competition in 1978; first developed a physically acceptable theory of faster-than-light particles; first derived a composition of extrema method in the Calculus of Variations; first quantitatively suggested that inflationary periods in the history of the universe were not needed; first proved Gödel's Theorem implies Nature must be quantum; provided a new alternative to the Higgs Mechanism, and Higgs particles, to generate masses; first showed how to resolve logical paradoxes including Gödel's Undecidability Theorem by developing Operator Logic and Quantum Operator Logic; first developed a quantitative harmonic oscillator-like model of the life cycle, and interactions, of civilizations; first showed how equations describing superorganisms also apply to civilizations. A recent book shows his theory applies successfully to the past 14 years of history and to *new* archaeological data on Andean and Mayan civilizations as well as Early Anatolian and Egyptian civilizations.

He first developed an axiomatic derivation of the forms of The Standard Model from geometry – space-time properties – The Extended Standard Model. It has a Dark Matter sector that approximates the ElectroWeak sector with Dark doublets and Dark gauge interactions. It also uses quantum coordinates to remove infinities that crop up in most interacting quantum field theories and additionally to remove the infinities that appear in the Big Bang and generate an inflationary growth of the universe. The Extended Standard Model has an ultra-high energy GUT (Grand Unified Theory) limit with a $U(4)\otimes U(4)$ symmetry; and can be united with gravitation to form a Theory of Everything. (See *Physics is Logic Part II.*)

Blaha has had a major impact on a succession of elementary particle theories: his Ph.D. thesis (1970), and papers, showed that quantum field theory calculations to all orders in ladder approximations could not give scaling deep inelastic electron-nucleon scattering. He later showed the eigenvalue equation for the fine structure constant α in Johnson-Baker-Willey QED had a zero at α = 1 not 1/137 by solving the Schwinger-Dyson equations to all orders in an approximation that agreed with exact results to $4^{th}$ order in α thus ending interest in this theory. In 1979 at Prof. Ken Johnson's (MIT) suggestion he calculated the proton-neutron mass difference in the MIT bag model and found the result had the wrong sign reducing interest in the bag model. These results all appear in Physical Review papers. In the 2000's he repeatedly pointed out the shortcomings of SuperString theory and showed that The Standard Model's form could be derived from space-time geometry by an extension of Lorentz transformations to faster than light transformations. This deeper space-time basis greatly increases the possibility that it is part of THE fundamental theory.Recently, Blaha showed that the Weak interactions differed significantly from the Strong, electromagnetic and gravitation interactions in important respects while these interactions had similar features, and suggested that ElectroWeak theory, which is essentially a glued union of the Weak interactions and Electromagnetism, possibly modulo unknown Higgs particle features, be replaced by a unified theory of the other interactions combined with a stand-alone Weak interaction theory. Blaha also showed that, if Charmonium calculations are taken seriously, the Strong interaction coupling constant is only a factor of five larger than the electromagnetic coupling constant, and thus Strong interaction perturbation theory would make sense and yield physically meaningful results.

In graduate school (1965-71) he wrote substantial papers in elementary particles and group theory: The Inelastic E- P Structure Functions in a Gluon Model. Phys. Lett. B40:501-502,1972; Deep-Inelastic E-P Structure Functions In A Ladder Model With Spin 1/2 Nucleons, Phys.Rev. D3:510-523,1971; Continuum Contributions To The Pion Radius, Phys. Rev. 178:2167-2169,1969; Character Analysis of U(N) and SU(N), J. Math. Phys. 10, 2156 (1969); and The Calculation of the Irreducible Characters of the Symmetric Group in Terms of the Compound Characters, (Published as Blaha's Lemma in D. E. Knuth's book: *The Art of Computer Programming Vols. 1 – 4*).

In the early 1980's Blaha was also a pioneer in the development of UNIX for financial, scientific and Internet applications: benchmarked UNIX versions showing that block size was critical for UNIX performance, developing financial modeling software, starting database benchmarking comparison studies, developing Internet-like UNIX networking (1982) and developing a hybrid shell programming technique (1982) that was a precursor to the PERL programming language. He was also the manager of the AT&T ten-year future products development database. His work helped lead to commercial UNIX on computers such as Sun Micros, IBM AIX minis, and Apple computers.

In the 1980's he pioneered the development of PC Desktop Publishing on laser printers. and was nominated for three "Awards for Technical Excellence" in 1987 by PC Magazine for PC software products that he designed and developed.

Recently he has developed a theory of Megaverses – actual universes of which our universe is one – with quantum particle-like properties based on the Wheeler-DeWitt equation of Quantum Gravity. He has developed a theory of a baryonic force, which had been conjectured many years ago, and estimated the strength of the force based on discrepancies in measurements of the gravitational constant G. This force, operative in 15-dimensinal space, can be used to escape from our universe in "uniships" which are the equivalent of the faster-than-light starships proposed in the author's earlier books. Thus travel to other universes, as well as to other stars is possible.

Blaha also considered the complexified Wheeler-DeWitt equation and showed that its limitation to real-valued coordinates and metrics generated a Cosmological Constant in the Einstein equations.

The author has also recently written a series of books on the serious problems of the United States and their solution as well as a book on the decline of Mankind that will follow from current social and genetic trends in Mankind.

In the past twelve years Dr. Blaha has written over 40 books on a wide range of topics. Some recent major works are: *From Asynchronous Logic to The Standard Model to Superflight to the Stars, All the Universe!, SuperCivilizations: Civilizations as Superorganisms, America's Future: an Islamic Surge, ISIS, al Qaeda, World Epidemics, Ukraine, Russia-China Pact, US Leadership Crisis, The Rises and Falls of Man – Destiny – 3000 AD: New Support for a Superorganism MACRO-THEORY of CIVILIZATIONS From CURRENT WORLD TRENDS and NEW Peruvian, Pre-Mayan, Mayan, Anatolian, and Early Egyptian Data, with a Projection to 3000 AD,* and *Mankind in Decline: Genetic Disasters, Human-Animal Hybrids, Overpopulation, Pollution, Global Warming, Food and Water Shortages, Desertification, Poverty, Rising Violence, Genocide, Epidemics, Wars, Leadership Failure.*

He has taught approximately 4,000 students in undergraduate, graduate, and postgraduate corporate education courses primarily in major universities, and large companies and government agencies.

The above paragraphs summarize much of his work over the past fifty years. This work is fully documented. He continues to engage in research and writing at Blaha Research.

www.ingramcontent.com/pod-product-compliance
Lightning Source LLC
Chambersburg PA
CBHW082113210326
41599CB00033B/6692